Leckie×Leckie

Scotland's leading educational publishers

National 3/4/5
APPLICATIONS OF MATHS

PRACTICE QUESTION BOOK

N3/4/5 APPLICATIONS OF MATHS
PRACTICE QUESTION BOOK

Craig Lowther • Mike Smith

© 2018 Leckie & Leckie Ltd

001/16042018

10 9 8 7 6 5 4 3 2 1

ISBN 9780008263553

Published by
Leckie & Leckie Ltd
An imprint of HarperCollinsPublishers
Westerhill Road, Bishopbriggs, Glasgow, G64 2QT
T: 0844 576 8126 F: 0844 576 8131
leckieandleckie@harpercollins.co.uk   www.leckieandleckie.co.uk

Commissioning Editor: Gillian Bowman
Managing Editor: Craig Balfour

Special thanks to
Jouve (layout and illustration); Ink Tank (cover design);
Jess White (copy edit); Nick Hamar (answer check);
Project One Publishing Solutions, Scotland (project management)

A CIP Catalogue record for this book is available from the British Library.

Printed and bound by CPI Group (UK) Ltd, Croydon, CR0 4YY

How to use this book

Welcome to Leckie and Leckie's *National 3/4/5 Applications of Maths Practice Question Book*. This book follows the structure of the Leckie and Leckie *National 3/4* and *National 5 Applications of Maths Student Books*, so is ideal to use alongside them, according to the course being studied. Questions have been written to provide practice for topics and concepts which have been identified as challenging for many students.

National qualification

Questions are clearly identified as N3, N4 or N5 to make it easy to find questions appropriate to each course.

Hints

Where appropriate, hints are provided to help give extra guidance and support.

Reasoning questions

Questions which require reasoning skills are marked with a ⚙ icon.

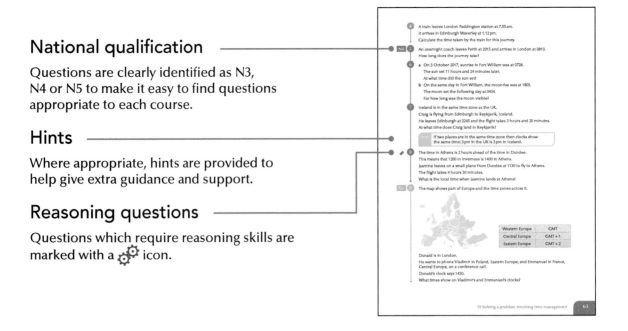

Use of calculators

Questions when you could use a calculator are marked with a 🖩 icon.
Questions when you should **not** use a calculator are marked with a 🗙 icon.

Examples

Examples with worked solutions provide support for particularly tricky concepts.

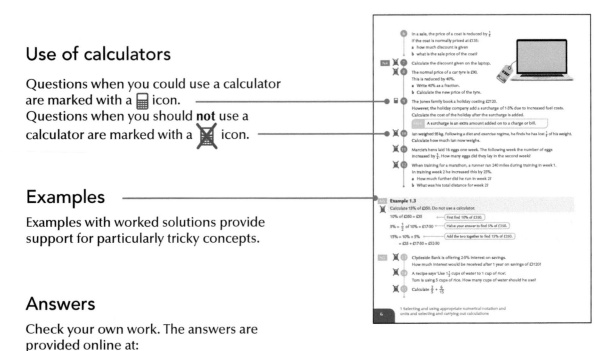

Answers

Check your own work. The answers are provided online at:

www.leckieandleckie.co.uk/page/Resources

1 Selecting and using appropriate numerical notation and units and selecting and carrying out calculations

Exercise 1A Applying the four operations to whole numbers and decimal fractions

> **Hint** In each question, you need to decide what type of calculation to do: add, subtract, multiply or divide.
>
> Remember to include the units in your answer.

N3

1 A bag of flour weighs 225 g.

Calculate the weight of 10 bags.

2 550 kg of gravel is being put into 10 sacks.

Calculate the weight of gravel in each sack.

3 Mike is fitting a kitchen worktop in a space which is 1840 mm long.

The worktop comes in a length of 2000 mm.

Find the length Mike needs to cut off in order for the worktop to fit.

4 James had £4·25 in his piggy-bank, Ellis had £6·42 and Hope had £5·12.

How much money did they have in total?

5 A bottle of orange juice costs £1·36.

Calculate the cost of 5 bottles.

6 Four friends go to the cinema. Each ticket costs £6·25.

Calculate the total cost of 4 tickets.

7 A plumber has a piece of pipe 2400 mm long.

She cuts it into 4 equal pieces. How long will each piece be?

8 **a** Colin is writing a cheque for a car repair.

The car repair cost £478.

Write this amount in words.

 b Gail is writing a cheque for a school trip for her daughter.

She writes 'three hundred and twenty-five pounds only'.

Write this amount in figures.

N4   **9** Johnny got £7·31 change when he bought a packet of bacon with a £10 note.

How much did the bacon cost?

10 A bowl of apples weighs 4·32 kg. A larger bowl with more apples weighs 5·83 kg.

4·32 kg

5·83 kg

  a Calculate the total weight of the two bowls of apples.
  b How much heavier is the larger bowl than the smaller one?

11 A carton of juice costs £1·57.

Find the cost of 8 cartons.

12 6 friends decided to split the cost of a meal in a restaurant equally.

The total bill was £61·08.

Calculate how much each would pay.

13 Mike bought 11 metres of cord to decorate his patio area.

Cord costs £8·23 per metre.

Calculate the total cost of the cord.

14 Julie has 36·72 cm of decorative tape. She cuts it into 8 equal pieces.

What is the length of each piece of tape?

N5

15 There are 1200 drawing pins in one box.

How many drawing pins will there be in 30 boxes?

16 There is 225 g of coffee in one jar.

Calculate the total weight of coffee in 20 of these jars.

17 A photocopier copies 12 840 pages in 60 minutes.

Calculate how many copies this is per minute.

> **Hint** 'Per' means 'for each', so 'per minute' means 'for each minute'.

18 The second prize in a prize draw was £5 200. The first prize was 11 times greater than the first prize.

What was the value of the first prize?

19 Votes in a local election were as follows:

  Smith 15 497   Bartlett 12 276   Edmond 9 122   Nimmo 16 032

  a Who won the election?

  b How many more votes than the second place candidate did the winning candidate get?

  c A total of 61 403 people were eligible to vote in the election.

  How many did not vote?

> **Hint** Eligible voters are all those who have registered to vote. This includes all those who did actually vote as well as those who chose not to vote.

1 Selecting and using appropriate numerical notation and units and selecting and carrying out calculations

20 In 2016, Robert's Tours had an income of £3·1 million.

In 2017 it had dropped to £2·65 million.

What was the decrease in the income for Robert's Tours?

Give your answer in thousands of pounds.

21 Mr Jones is a manager in a small company. He earns £33 450 per year.

Ms Clark is Chief Executive of the company. She earns 7 times more than Mr Jones.

How much does Ms Clark earn?

22 In a motor cycle race, Enzo averaged 43·27 seconds per lap.

What was his time for 20 laps?

23 The speed of sound, Mach 1, is about 761·26 mph.

In 2004, a NASA X-43A aircraft flew at 9 times that speed (Mach 9).

At what speed, in miles per hour, was the NASA X-43A flying?

24 A company paid a total bonus of £67 200.

This was shared equally among its 800 workers.

Calculate the share each worker received.

Exercise 1B Rounding answers

N3 **Example 1.1**

Round 3·141 592 to 3 decimal places.

3·141592 — | Underline the digits you want to round to. |

Hint | When rounding to a given number of places, look at the digit on the right of the one you are rounding up or down.

3·141592 — | Look at the next digit. Is it 5 or more? In this example, yes, so you need to round **up**. |

If it is 5 or more, round **up** the digit you are rounding, that is, increase it by 1.

3·142 (to 3 d.p.) — | Increase the digit in the third decimal place by 1, so change 1 to 2. |

If it is less than 5, do not change the digit you are rounding.

N3 **1** The distance, by car, from Edinburgh to Glasgow is 72 km.

Round this distance to the nearest 10 km.

2 Loch Ness is 36·3 km long.

Write this to the nearest 10 km.

3 The height of the Eiffel Tower is 324 m.

Write this to the nearest 100 m.

4 The diameter of the Sun is about 1·3914 million km.

Round this to 2 decimal places.

5 The average cell in the human body has a diameter of 0·051 mm.

Write this to 1 decimal place.

6 Sunita says there are 170 pupils in her year at school.

If she has rounded this to the nearest 10, how many pupils could there be in her year at school? Write all of the possible answers.

7 Tomas divides 47·28 by 15 on his calculator and gets the answer 3·152. The question says 'Write your answer to 1 decimal place'.

What should Tomas write as his answer?

8 7 friends equally share a prize win of £319 between them.

How much does each friend get? Round your answer to 2 decimal places.

N4 **Example 1.2**

Round the following to 3 significant figures.

a 14·683   b 0·0036147

a 1<u>4·6</u>83 ●————— Identify the first non-zero digit. Then underline the number of digits you wish to round to.

14·7 (3 s.f.) ●————— Round using the same rules as before: if the next digit is 5 or more, increase the digit you are rounding by 1; if the next digit is less than 5 leave the digit as it is.

b 0·00<u>361</u>47

0·00361 (3 s.f.)

N4 **9** A symbol you may see in mathematics is π. (This is called 'pi'.)

π written to 5 decimal places is 3·14159.

Round 3·14159 to:

a 2 decimal places   b 2 significant figures.

10 The diameter of the Earth is 12742 km.

Write this to 2 significant figures.

11 The circumference of the Sun is about 2713406 miles.

Write this to 1 significant figure.

12 Jakub multiplies 31·56 by 17·1 and gets the answer 539·676.

a The question says 'Write your answer to 2 decimal places'. What should Jakub write as his answer?

b If the question had said 'Write your answer to 2 significant figures', what should Jakub have written as his answer?

13 Three friends weighed themselves, and the results were as follows:

   Walcott, 84·58 kg   Ruuskanen, 84·12 kg   Vesely, 83·34 kg

Write each weight correct to 3 significant figures.

14 Anne recorded the temperature of water in a cup. The thermometer showed 65·94 °C. Write this correct to 1 decimal place.

15 The formula used to calculate the time it takes for a diver, from a diving board, to hit the water is $t = \sqrt{\dfrac{h}{5}}$ where t is the time, in seconds, and h is the height, in metres, of the diving board. Ola dives from the 3 m diving board.

Calculate the time it takes for Ola to hit the water. Give your answer to 3 significant figures.

16 A number often used in mathematics is e, which is approximately 2·718 28.

   a Write the number e to:

   i 1 decimal place   **ii** 2 decimal places   **iii** 3 decimal places.

   b Write the number e to:

   i 1 significant figure   **ii** 2 significant figures   **iii** 3 significant figures.

17 Robyn's circular table has a diameter of 1·8 m. She calculates the area to be 2·5446 m². Write this area correct to 2 significant figures.

Exercise 1C Working with fractions and percentages

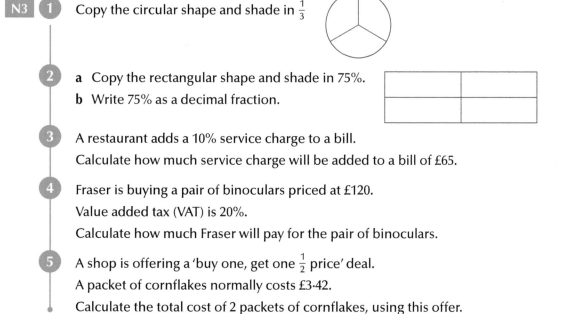

1 Copy the circular shape and shade in $\frac{1}{3}$

2 **a** Copy the rectangular shape and shade in 75%.

   b Write 75% as a decimal fraction.

3 A restaurant adds a 10% service charge to a bill.

Calculate how much service charge will be added to a bill of £65.

4 Fraser is buying a pair of binoculars priced at £120.

Value added tax (VAT) is 20%.

Calculate how much Fraser will pay for the pair of binoculars.

5 A shop is offering a 'buy one, get one $\frac{1}{2}$ price' deal.

A packet of cornflakes normally costs £3·42.

Calculate the total cost of 2 packets of cornflakes, using this offer.

6 In a sale, the price of a coat is reduced by $\frac{1}{5}$
If the coat is normally priced at £135:
  a how much discount is given
  b what is the sale price of the coat?

N4 **7** Calculate the discount given on the laptop.

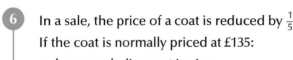

8 The normal price of a car tyre is £90.
This is reduced by 40%.
  a Write 40% as a fraction.
  b Calculate the new price of the tyre.

9 The Jones family book a holiday costing £2120.
However, the holiday company add a surcharge of 1·5% due to increased fuel costs.
Calculate the cost of the holiday after the surcharge is added.

> **Hint** A surcharge is an extra amount added on to a charge or bill.

10 Ian weighed 95 kg. Following a diet and exercise regime, he finds he has lost $\frac{1}{5}$ of his weight. Calculate how much Ian now weighs.

11 Marcie's hens laid 16 eggs one week. The following week the number of eggs increased by $\frac{3}{4}$. How many eggs did they lay in the second week?

12 When training for a marathon, a runner ran 240 miles during training in week 1.
In training week 2 he increased this by 25%.
  a How much further did he run in week 2?
  b What was his total distance for week 2?

N5
Example 1.3

Calculate 15% of £350. Do not use a calculator.

10% of £350 = £35   ○——(First find 10% of £350.)

$5\% = \frac{1}{2}$ of 10% = £17·50   ○——(Halve your answer to find 5% of £350.)

15% = 10% + 5%   ○——(Add the two together to find 15% of £350.)

    = £35 + £17·50 = £52·50

N5 **13** Clydeside Bank is offering 2·5% interest on savings.
How much interest would be received after 1 year on savings of £3120?

14 A recipe says 'Use $1\frac{1}{2}$ cups of water to 1 cup of rice'.
Tom is using 5 cups of rice. How many cups of water should he use?

15 Calculate $\frac{2}{5} + \frac{6}{15}$

1 Selecting and using appropriate numerical notation and
units and selecting and carrying out calculations

16 A 250 g pack of cereal contains 9 g of protein and 4 g of vitamins.

    a What percentage of the weight of the cereal is protein?

    b What percentage of the weight of the cereal is vitamins?

17 Mark recently sat three tests. He got $\frac{34}{50}$ for physics, $\frac{14}{18}$ for mathematics and $\frac{59}{80}$ for biology.

    Rank his tests in order, from best to poorest result.

18 Sajid left £2500 in the bank for 4 years at a compound interest rate of 3·2%.

    a Calculate how much he had in the bank after 4 years.

    b How much interest did Sajid receive?

19 Amna bought a car for £15 500. It depreciates in value at a rate of 17% per year.

    Calculate the value of Amna's car after 5 years.

20 A company replaces its machinery when the machine's value drops below half its original value.

    A machine is bought for £5000. It depreciates in value at a rate of 15% per year.

    After how many years should the company replace the machine?

21 The manufacturer of a cleaning spray claims that, when sprayed on a kitchen surface, it will kill at least 80% of bacteria within 10 seconds.

    A kitchen surface has 18 000 bacteria on it when the cleaning spray is sprayed onto it.

    10 seconds later there are 3750 live bacteria on the surface.

    Is the manufacturer's claim valid? Use your working to justify your answer.

22 Julie thinks that if she invests her money at 6·4% compound interest, she will double her money in 10 years.

    Is Julie correct? Use your working to justify your answer.

Exercise 1D Calculating speed, distance and time

N4 **Example 1.4**

A car travelled 206 km in 2 hours 15 minutes.

What was its average speed? Give your answer to 1 decimal place.

$15 \div 60 = 0{\cdot}25$ ⟶ First, find the time in hours by converting 15 minutes to a decimal fraction of an hour.

So 2 h 15 m = 2·25 h

$\text{speed} = \dfrac{\text{distance}}{\text{time}}$ or $s = d \div t$

$\quad = \dfrac{206}{2{\cdot}25}$

$\quad = 91{\cdot}5555... \, \text{kph}$

$\quad = 91{\cdot}6 \, \text{km/h} \, (1 \, \text{d.p.})$ ⟵ Round to 1 decimal place.

1 A plane leaves Edinburgh Airport at 1050 and arrives at Malaga Airport at 0215.
How long does the flight take?

2 Part of a bus timetable is shown.
If Peter wishes to be in Rosewell for 1415, which bus, from Eskbank, should he take?

Dalkeith	1307	1327	1347	1407	1427	1447
Eskbank	1312	1332	1352	1412	1432	1452
Bonnyrigg	1320	–	1400	–	1440	1500
Lasswade	1328	1348	–	1428	–	1508
Rosewell	1337	1357	1417	1437	1457	1517

3 Maisie is riding her bike at 15 mph.
How far will she travel in 6 hours if she rides at the same speed?

4 Margaret is driving at 40 mph. Her journey is a distance of 240 miles.
How long will it take Margaret to complete the journey if she drives at the same speed?

5 A plane left Paris at 1045 and arrived in Edinburgh at 1:45 pm.
The plane flew at 295 kph.
How far did the plane travel?

N5 **6** Calum is driving from Edinburgh to Dover, a distance of 470 miles, to catch a ferry to France.
He needs to be at the ferry for 0830.
He expects to average 50 mph.
He will have a stop of 45 minutes for a coffee and another of 25 minutes to re-fuel.
At what time should Calum leave Edinburgh in order to arrive at the ferry on time?

7 Rory competed in a car rally event.
He completed the 132 km course in 1 hour 42 minutes.
What was his average speed?
Give your answer to 1 decimal place.

8 A dinghy is sailing at an average speed of 15 km/h for 36 minutes.
How far does it travel?

9 Jack and Jill are driving their motorhome at an average speed of 65 km/h.
How far will they travel in 3 hours 12 minutes?

10 Grigor is steering a narrow boat along a canal which is 42 km long.
He can average a speed of 4·8 km/h.
How long will it take him to travel the length of the canal?
Give your answer in hours and minutes.

Exercise 1E Working with perimeter, circumference, area and volume

N4 **1** Calculate the perimeter of each of these floor plans.

a

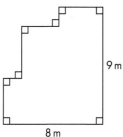

9 m

8 m

b

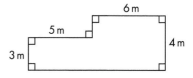

6 m

5 m

3 m

4 m

2 Calculate the area of this floor.

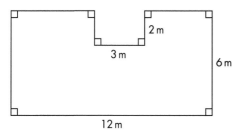

2 m

3 m

6 m

12 m

3 Calculate the volume of this box.

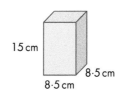

15 cm

8·5 cm

8·5 cm

N5 **4** Calculate the area of this garden.

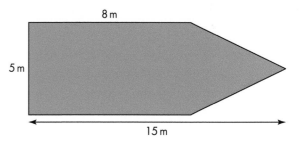

8 m

5 m

15 m

5 A rectangular metal plate has 4 holes cut out of it, as shown in the diagram.

The holes are circles of diameter 9 mm.

Calculate the area of remaining metal.

Give your answer to 2 decimal places.

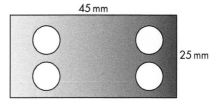

45 mm

25 mm

> **Hint** The area A of a circle is found using the formula $A = \pi r^2$ where r is the radius.

6 The diagram shows a farmer's field.

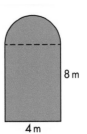

8 m

4 m

  a Calculate how much fencing will be
  needed to go around the field.

  b Calculate the area of the field.

Give your answers to 3 significant figures.

> Hint The circumference C of a circle
> is found using the formula
> $C = \pi d$ where d is the diameter.

7 Jane has a jewellery box which is a hexagonal prism (left). The hexagonal base of the
box is made of 6 equal triangles, as shown (right).

> Hint Calculate the area
> of one triangle,
> then multiply by the
> number of triangles
> to find the area of
> base.

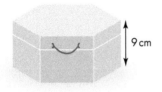

9 cm

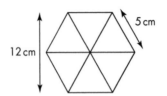

5 cm

12 cm

The depth of the box is 9 cm. Calculate the volume of the box.

8 Doggo dog food is sold in cylindrical tins
with dimensions as shown.

What is the volume of the tin?

Give your answer to 3 decimal places.

6 cm

12 cm

> Hint Remember the formula for finding the volume V of a
> cylinder is $V = \pi r^2 h$ where r is the radius and h is the height.

9 A chocolate box is in the shape of a triangular prism,
with dimensions as shown.

Calculate the volume of the box.

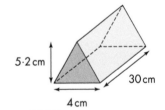

5·2 cm

4 cm

30 cm

10 A toy lunch box is in the shape of a cuboid with a
half-cylinder lid, as shown.

Calculate the volume of the lunch box.

8 cm

15 cm

8 cm

Exercise 1F Calculating ratio

N4 **1** Frank and Ben share £480 in the ratio 1:3.

How much money do they each get?

 2 Freya and Eoin share 42 colouring pencils in the ratio 4:3.

How many pencils do they each get?

3 Beth and Sue share 40 sweets in the ratio 11:9.

How many more sweets does Beth get than Sue?

N5 **4** Tom, Dick and Harry share £84 000 in the ratio 2:3:7.

How much money do they each get?

5 A shop sold 60 designer dresses and 24 pairs of gold earrings one weekend.

Express the ratio of dresses to earrings sold.

Simplify as far as possible.

6 Sharon and Tanya work for the Post Office delivering letters.

One day they delivered 1 012 letters in the ratio 4:7.

How many letters did each deliver?

7 A copper pipe is 2000 mm long.

It is cut into three sections in the ratio 12:7:6.

How long is each section?

8 Teachers in the Maths Department at Balvennie High School win a lottery.

Being maths teachers, they decide to share the winnings in proportion to how much each teacher pays per week.

They each pay the following amounts per week:

Mr Smith	£2·00
Ms Flanagan	£1·50
Mr Beard	£0·50
Mr McKay	£3·00
Ms Cameron	£2·00

Mr McKay's share of the win is £204 000.

Calculate how much the teachers won in total.

Exercise 1G Calculating direct and indirect proportion 🖩

4 **Example 1.5**

8 litres of turf seed cost £34·80.

What would 11 litres cost?

Litres   Cost (£)

8   →   34·80

1   →   34·80 ÷ 8 ⟵ Calculate the cost for 1 litre.

= £4·35 per litre

11   →   4·35 × 11 ⟵ Calculate the cost for 11 litres.

= £47·85

11 litres would cost £47·85.

Example 1.6

It took 3 people 8 hours to lay a lawn.

How many hours would it take 5 people?

People		Hours
3	→	8
1	→	8 × 3
		= 24
5		24 ÷ 5
		= 4·8 hours

8 × 3 ── Work out the number of hours needed to complete the task.

24 ÷ 5 ── Share the hours between the number of people.

It would take 5 people 4·8 hours.

N4 **1** When Clara was going on holiday she exchanged £50 for €60.

How many euros (€) would she get for £70?

2 7 birthday cakes cost £56.

How much would 9 birthday cakes cost?

3 A space shuttle took 125 hours to orbit the Earth 5 times.

How long would it take to orbit the Earth 7 times?

N5 **4** Bryan has a recipe to make 16 gingerbread men.

Calculate how much of each ingredient is needed to make 24 gingerbread men.

Ingredient	Amount for 16 gingerbread men
flour	180 g
ginger	40 g
butter	110 g
sugar	30 g

5 225 g of flour are needed to make 9 cakes.

Maria has 480 g of flour. She thinks she has enough to make 20 cakes.

Is Maria correct? You must show all your working.

6 A piece of wire, 25 cm long, weighs 8 g.

Another piece of the same wire is 30 cm long.

How much does it weigh?

7 16 workers can clean up a sports stadium in 2 hours.

If only 10 workers turn up, how long will it take to clean up the stadium?

Give your answer in hours and minutes.

8 It takes 6 men 12 days to erect scaffolding around an office block.

If the scaffolding has to be put up in 9 days, calculate how many extra men will be required.

1 Selecting and using appropriate numerical notation and
units and selecting and carrying out calculations

2 Recording measurements using a scale on an instrument

Exercise 2A Measure to the nearest marked, minor unnumbered division on a scale

N3 **1** Write down the length, to the nearest centimetre, of each object shown.

a

b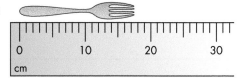

2 Write down the weights, to the nearest kilogram, shown on these scales.

a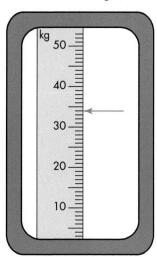

b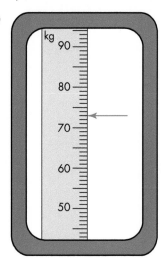

3 Write down the temperatures, to the nearest degree Celsius, shown on these thermometers.

a

b

4 Write down the temperatures, to the nearest 10 degrees C, indicated by the arrows on these oven controls.

a

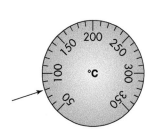

b

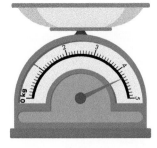

5 Write down the weight shown on these scales.

a

b

6 Write down the temperature shown on these thermometers.

a
b
c
d

7 What is the volume of liquid in these measuring jugs?

a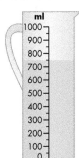
b
c
d

8 Write down the length of this millipede, including the antennae.

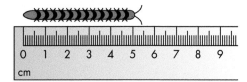

2 Recording measurements using a scale on an instrument

9 What size is the angle marked on these protractors?

a

b

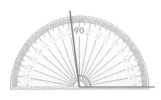

c

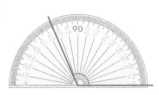

d

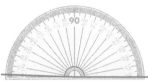

10 What is the size of angle marked on these angle measures?

a

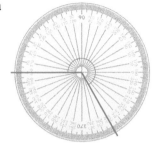

b

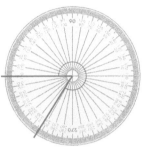

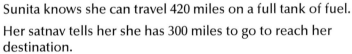

11 Sunita knows she can travel 420 miles on a full tank of fuel.

Her satnav tells her she has 300 miles to go to reach her destination.

Does Sunita have enough fuel in her tank to complete her journey?

Use your working to justify your answer.

12 Chris is checking his racing bike before setting off on a race.

He wants his tyres to be at 80 psi.

How much does he need to increase the pressure by to get his tyre to the correct pressure?

13 Alison is going on holiday and is weighing her luggage.

The airline will charge extra if the total weight of bags is more than 30 kg.

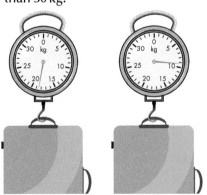

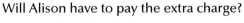

Will Alison have to pay the extra charge?

You must justify your answer.

2 Recording measurements using a scale on an instrument

3 Interpreting measurements and the results of calculations to make decisions

Exercise 3A Identifying relevant measurements and using results of calculations to make a decision

N3 **1** Flora has a worktop which is 3000 mm in length.

She cuts 3 pieces, each 680 mm, from it.

a Estimate how much of the worktop is left.

b Calculate how much worktop is left.

2 Write down the following objects in order, starting with the smallest volume.

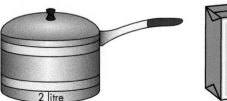 

3 Patrick is checking the weight of his suitcases before going on holiday.

He puts them on the bathroom scales to check. The diagrams below show the weight of the suitcases.

The airline states 'Total weight of suitcases must not exceed 21 kg'.

Will Patrick be allowed to take his suitcases on the plane?

N4 **4** Gena lives 1·7 km from school.

She walks there and back 5 times per week.

How many metres does Gena walk in total each week to school and back?

5 David Rudisha, of Kenya, holds the world record for the 800 m track running event.

His time is shown on the stopwatch on the left.

Sebastian Coe, of Great Britain, ran the 800 m in the time shown on the stopwatch on the right.

How much faster was Rudisha?

6
a Calculate the volume of this box in cm³.

b How many litres is this?

Hint 1 litre = 1 000 cm³

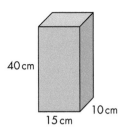

40 cm

10 cm

15 cm

7
The world record for throwing the discus was set by Jurgen Schult in 1986, at 74·08 m.

The Olympic record was set by Virgilijus Alekna in 2004, at 69·98 m.

How much further did Schult throw the discus?

Give your answer in centimetres.

8 Carol has bought 10 litres of juice for a party.

She has 28 guests coming and estimates that each will drink 400 ml of juice

Does Carol have enough juice?

Use your working to justify your answer.

N5 **9** The table helps you to convert weights from stones to kilograms.

a Jack weighs 14 st.

How many kilograms is this?

b Joe weighs 76 kg.

Approximately how many stones is this?

c Body mass index (BMI) is an approximate measurement used to indicate whether a person's weight is healthy. The formula for calculating this is:

$$BMI = \frac{weight\ (kg)}{height\ (m)^2}$$

Jack is 1·72 m tall.

What is Jack's BMI?

Give your answer to 1 decimal place.

Stones	Kilograms
1 st	6·35 kg
2 st	12·70 kg
3 st	19·05 kg
4 st	25·40 kg
5 st	31·75 kg
6 st	38·10 kg
7 st	44·45 kg
8 st	50·80 kg
9 st	57·15 kg
10 st	63·50 kg

10 Fraser is driving along an autoroute in France.
He sees a speed limit sign of 120 kilometres per hour (km/h).

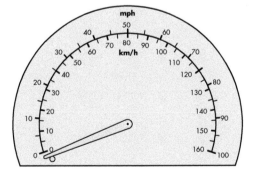

a Use the speedometer to change 120 km/h to miles per hour (mph).

b Fraser knows that in Britain, in a built-up area, the speed limit is 30 mph.

What would this limit be in kilometres per hour?

11 Colin weighs 145 pounds (lb).

His bike has 700 × 25c tyres fitted.

The tyre gauge shows the pressure in Colin's front tyre.

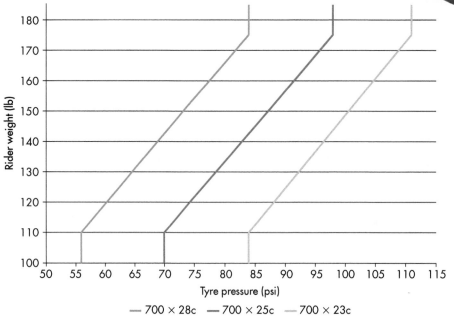

Tyre pressure (psi)

— 700 × 28c — 700 × 25c — 700 × 23c

State any adjustment Colin should make to the air pressure in his front tyre.

Your statement should include whether to increase or decrease the tyre pressure and by how much.

12

a Normal body temperature is 36·8 °C.

What is this in degrees Fahrenheit?

b The freezing point of water is 32 °F.

What is this in degrees Celsius?

c The highest recorded temperature in Scotland was 32·9 °C.

This was recorded in Greycrook, Scottish Borders, on 9 August 2003.

The average summer temperature in Scotland is 59 °F.

What is the difference in these temperatures?

4 Justifying decisions by using the results of measurements and calculations

Exercise 4A Using evidence from the results of calculations to justify decisions

N3 **1** Bella has £5. She wants to buy potatoes.

What other two items could she buy?

You must show all your working.

2 Julie buys three beauty products from a store. The products cost £11·45, £17·23 and £21·14.

Julie hands over three £20 notes.

The shopkeeper gives Julie change of £9·18.

Has the shopkeeper made a mistake?

Use your working to explain your answer.

3 A forklift truck can safely carry a weight of 1000 kg.

A container full of metal equipment weighs 128 kg.

Can the forklift truck safely carry 8 of these containers at one time?

Use your working to explain your answer.

N4 **4** This juice carton says it holds 1 litre of apple juice.

Does the carton hold at least the stated amount?

Use your working to explain your answer.

5 Rajid takes the bus to work and back, 5 days each week.

A ticket for a single journey costs £1·70.

He could buy a monthly (4 week) season ticket for £65·50.

How much would Rajid save if he buys the season ticket?

6 Wiktoria is looking to hire a car for 7 days. She sees these two adverts.

Cars 4 U
car hire
£24·50 per day
no mileage charge

We R Cars
car hire
£17·90 per day plus
£0·52 per mile over
100 miles

She estimates she will cover 200 miles during the 7 days.

Which company would be cheaper for Wiktoria?

7 Peter weighed himself on 1 June.

The scales showed he weighed 97 kg.

He is going on holiday on 2 November and wants to weigh 80 kg by then.

On 1 June, Peter hired Colin to develop a programme of exercise and diet to help him lose weight.

Colin reckons his programme will help Peter to reduce his weight by 4% per month.

If Colin's programme is successful, will Peter reach his target weight by the time he sets off on holiday?

Use your working to justify your answer.

8 Building regulations state that the gradient of a ramp giving access to a ground floor apartment must be less than $\frac{1}{14}$

Does this ramp meet the building regulations?

Use your working to explain your answer.

28 cm

3·2 m

9 The 'Box It Right' company are designing two storage boxes.

Each box needs to hold 20 litres.

Their designer has suggested two versions, Box A and Box B, as shown.

a Do both boxes hold at least 20 l?

b The company will choose the design which uses the least amount of material to make.

Which box should the company choose?

Use your working to justify your answer.

Box A

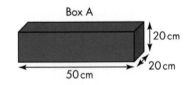

20 cm
50 cm
20 cm

Box B

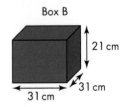

21 cm
31 cm
31 cm

Hint | Calculate the total **surface area** of each of the boxes. Surface area of a cuboid is the total area of all the rectangular faces.

10 Ellis knows that she can visit her mother and return home, by car, in 3 hours if she drives at an average speed of 60 mph.

To improve road safety the government has reduced the speed limit on the road to 50 mph.

How long should Ellis now allow for her journey?

Give your answer in hours and minutes.

5 Analysing a financial position using budget information

Exercise 5A Budgeting and planning, balancing income and expenditure

N3 **1** Calculate the total cost of this bill. Give your answer in pounds and pence.

1 packet of pasta	£2·35
2 tins of tomatoes	45p each
3 cucumbers	37p each
2 packets of cereal	£2·42 each
Total	

2 Mr and Mrs Chisholm and their 3 children are booking their summer holiday.

The cost for 1 adult is £549. The cost for 1 child is £384.

a Find the total cost of the holiday for the 2 adults and 3 children.

b The Chisholm family save £375 per month towards their holiday.

How many months will it take them to save up for the holiday?

N4 **3** Here is Lisa's budget sheet for January.

Income	£	Expenditure	£
Wage	950·50	Rent	275·00
Share dividend	82·47	Electricity / gas	67·43
		Food	257·62
		Credit card repayment	23·87
		Car loan	93·50
		Car insurance and costs	85·38
		TV licence	17·00
		Mobile phone / internet	35·00
		Entertainment	60·00
Surplus / deficit spend: Income – Expenditure			

a Does Lisa have a surplus or a deficit?

b Calculate her surplus or deficit.

c In February, Lisa did not receive a share dividend.

If her expenditure stayed the same, what impact would this have on Lisa's budget?

4 Last year, the Green family had an average monthly income of £2400.

Their average monthly expenditure was £2225.

This year the Green family income rose by 0·5%.

Their expenditure rose by 8%.

Calculate the new monthly surplus.

5   Jemma and Ash make a list of expected costs and income for their charity dance.

Costs	£	Income	£
Hire of hall	110	Tickets (100)	
Juice	20	Raffle	80
Food	40		
Music	20		
TOTAL COSTS		TOTAL INCOME	

Jemma and Ash think that 100 people will attend and that they will raise £80 by selling the raffle tickets. They want to raise £500 for charity.

Calculate how much they will have to charge for each ticket to raise this amount.

6   Serena is going on holiday and has to pay the final balance in 12 weeks' time.

The 'all inclusive' cost of the holiday is £1100.

She earns £9·60 per hour, and works for 35 hours each week.

She records her outgoings to help calculate if she will save enough for the holiday. Any money left after her outgoings will go into her holiday savings fund.

Item	Income (£)	Outgoings (£)
Pay		–
Rent	–	84
Food	–	48
Utilities	–	51
Transport	–	12
Socialising	–	20
TV/phone/broadband	–	22
HOLIDAY SAVINGS FUND		

Copy and complete the table to answer the following.

a   How much will go into Serena's holiday savings fund each week?

b   Will she be able to save enough for her holiday?

c   If Serena paid for the holiday by credit card, there would be a 2·5% added charge on the cost. How much extra would she pay if paying by credit card?

7   Finlay writes down his monthly budget.

Income	Expenditure	
£1600	Mortgage	£615
	Council tax	£227
	Food	£345
	Transport	£120
	Bills	£285

a   i   Does Finlay have a surplus or a deficit?   ii   How much surplus or deficit?

b   Finlay's transport costs rise by 10%. How much will his surplus or deficit now be?

6 Analysing and interpreting factors affecting income

Exercise 6A Income and deductions for different personal circumstances and career choices 🖩

Example 6.1

Jamie earns £8·20 per hour for each hour worked, Monday to Friday.

If he works on a Saturday he gets paid overtime at time-and-a-half.

Here is Jamie's timesheet for the first week in February.

What will Jamie's gross pay be for this week?

Week 1	Times	Hours
Monday	0800–1200, 1300–1600	7
Tuesday	0800–1200, 1300–1600	7
Wednesday	0800–1200	4
Thursday	0800–1200, 1300–1600	7
Friday	0800–1200, 1300–1600	7
Saturday	1000–1300	3

Monday–Friday: 32 hours × £8·20 = £262·40

Saturday: $1\frac{1}{2}$ × £8·20 = £8·20 + £4·10 = £12·30

       3 hours × £12·30 = £36·90

Total pay: £262·40 + £36·90 = £299·10

> Time-and-a-half means $1\frac{1}{2}$× hourly pay.

N3

1 Caroline works in a clothes shop and earns £8·15 per hour.

How much will Caroline earn when she works for 15 hours?

2 Imran sees these three job adverts in the paper.

Which one would give him the highest gross pay?

Carpet fitter	*Shop fitter*	Gas fitter
Salary £27 000 per annum	*£2215 per month*	£512 per week

3 Copy this payslip and complete it to find Neil's take home pay.

Neil Campbell	Period 5	April 2017	NI 682077B
Basic pay	Overtime	Bonus	Gross pay
2015·00	240·00	–	£
Tax	NI	Union fees	Deductions
260·00	180·00	12·50	£
Net pay			£

4 Cameron is working during the summer holidays cleaning caravans on a caravan site.

He gets £23·50 for every caravan he cleans.

If he cleans more than 5 caravans he gets a bonus of £50.

How much will Cameron get on a day when he cleans 7 caravans?

5 Pat works in a library and gets paid £8·20 per hour.

As the library is open at weekends, Pat gets an additional 5% per hour 'unsocial hours' pay for those hours worked at the weekend.

How much will Pat earn in a week in which she works 20 hours from Monday to Friday and 8 hours at the weekend?

6 Lucy has an annual salary of £23 150. Her personal allowance is £11 500.

a Calculate her taxable income.

b She pays 20% of her taxable income in income tax.

How much income tax will she pay annually?

Example 6.2

This table shows how much employers deduct for National Insurance (NI) from employees' pay for the 2017/18 tax year.

Employee group	Category letter	£113 to £157 a week (£490 to £680 per month)	£157·01 to £866 a week (£680·01 to £3750 per month)	Over £866 a week (£3750 per month)
All employees apart from those in groups B, C, J, H, M and Z	A	0%	12%	2%
Married women and widows entitled to pay reduced NI	B	0%	5·85%	2%
Employees over the State Pension age	C	N/A	N/A	N/A
Apprentice under 25	H	0%	12%	2%
Employees who can defer NI because they're already paying it in another job	J	0%	2%	2%
Employees under 21	M	0%	12%	2%
Employees under 21 who can defer NI because they're already paying it in another job	Z	0%	2%	2%

Source: www.gov.uk

Aisha is in category A and earns £1000 per week. What will she pay in National Insurance?

Nothing on the first £157

12% on her earnings between £157·01 and £866   = £85·08

2% on her earnings over £866   = £2·68

Total National Insurance payment   = £87·76

> **Hint** See Exercise 1C for a reminder about working with percentages.

7 Use the table on page 24 to calculate the National Insurance contribution of the following people.

a   Bill, an employee in category A who earns £750 per week.

b   Sandra, an apprentice under 25 who earns £1250 per month.

c   Erin, who is a 19-year-old employee, earning £155 per week.

d   Yolande, a widow who earns £1100 per week.

8 Gordon sells cars and earns £27 000 per annum.

His personal allowance is £11 500 per annum.

Gordon pays tax at the basic rate of 20% of taxable income.

He pays National Insurance at a rate of 12% on earnings over £8160.

Use this information to complete Gordon's payslip.

Period: March		Name: Gordon McKenzie			
Income (£)		**Deductions (£)**			
Basic	2250·00	Income tax	a	Gross pay	2250·00
Bonus	0·00	NI	b	Deductions	c
Total	2250·00	Total deductions	c	NET PAY	d

9 Stacey makes fused-glass jewellery.

If she works for 5 hours a day, she can make 8 pieces of jewellery.

A designer shop wants to order 32 pieces of jewellery and gives Stacey two payment options.

A   A flat fee of £10·75 per hour for the time it takes to make the pieces.

B   An upfront payment of £50 and £5 per item of jewellery.

Which option would give Stacey more money?

Use your working to justify your answer.

10 Dawn is a Chief Inspector in the police and earns £58 000 per year.

Her personal allowance is £11 500.

She pays the basic rate of tax of 20% on taxable income up to £45 000.

She pays the higher rate of tax of 40% on taxable income over £45 000.

What is Dawn's income tax payment for the year?

7 Determining the best deal

Exercise 7A Comparing at least three products, given three pieces of information on each 🖩

N3 **1** Caroline is learning to drive and sees this offer for driving lessons:

- 1 lesson: £24·99
- Special offer: 6 lessons for £140

How much will Caroline save if she takes the special offer?

2 A shop is advertising ring-binders for sale.

One ring-binder 90p   Pack of 5 ring-binders £4·25

Which is the better deal?

3 Two bottles of shampoo are for sale in a shop.

Which is the better value for money?

4 Amanda is shopping online for a TV set.

She sees these two adverts.

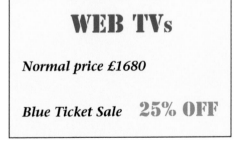

Which would give Amanda the better deal, that is, the cheaper TV set?

Example 7.1

A shop is selling the drink Kola in three different sized containers.
Which container is the best value for money?

2 l bottle: £2·35 ÷ 2 = £1·175 per litre

1·5 l bottle: £1·53 ÷ 1·5 = £1·02 per litre

500 ml can: £0·62 ÷ 0·5 = £1·24 per litre

> To compare the value offered by each container, work out the cost per litre.

So the best value is the 1·5 l bottle as it is the cheapest per litre.

5 A shop is selling bottles of conditioner as shown.

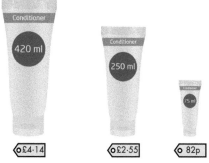

Which is the best value for money?

6 Ladl has an offer on toothpaste.

- 50 ml tube costing £1·40

- 75 ml tube costing £1·80

- Buy 2 × 50 ml tubes and get a third tube half-price.

Which is the best deal?

Fred wants to buy orange juice.

His local supermarket has the following options.

Pack of four
1 litre cartons
£5·65

1 litre carton
£1·42

Pack of 3
330 ml cartons
£1·38

Which offers the best value for money?

Use your working to justify your answer.

8   Nazeem wants to buy a train ticket to go to Manchester.

He uses a comparison site to check prices and sees these three offers.

RAILTRACK

Basic ticket price £90
Credit card charge
2·25% of ticket price
Booking fee £1·25

CHEAPLINE

Basic ticket price £90
Credit card charge
2% of ticket price
Booking fee £1·80

LOWRAILFARE

Basic ticket price £90
Credit card charge
2·8% of ticket price
Booking fee £nil

Nazeem will pay for the ticket using a credit card.

Which company will give Nazeem the best value?

You must show all your working and justify your answer.

9   Morag wants to buy re-writable CDs to record some of her own music tracks onto a CD.

She sees these offers.

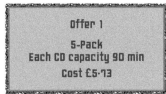

Offer 1

5-Pack
Each CD capacity 90 min
Cost £5·73

Offer 2

10-Pack
Each CD capacity 80 min
Cost £10·30

Offer 3

12-Pack
Each CD capacity 75 min
Cost £11·50

Which offer would give her best value?

You must show your working and justify your answer.

8 Converting between several currencies

Exercise 8A Converting between currencies 🖩

Use this conversion table to help you answer the questions in this exercise.

	GBP £	US $	Euro €	Swiss franc CHF	Chinese yuan CNY	South African rand ZAR	Emirati dirham AED
GBP £	1	1·68	1·24	1·48	10·42	17·18	6·15
US $	0·60	1	0·84	0·96	6·53	12·96	3·67
Euro €	0·81	1·19	1	1·14	7·82	15·49	4·39
Swiss franc CHF	0·68	1·04	0·88	1	6·83	13·54	3·84
Chinese yuan CNY	0·10	0·15	0·13	0·15	1	1·98	0·56
South African rand ZAR	0·06	0·08	0·06	0·07	0·51	1	0·28
Emirati dirham AED	0·16	0·27	0·23	0·26	1·79	3·57	1

Example 8.1

James exchanges £200 for Emirati dirham. How many dirham does James get?

£1 = 6·15 AED •————(Look down the left-hand column to find GBP £, then along to the Emirati dirham column to find how many dirham James will get for £1.

£200 = 6·15 × 200 •————(Multiply the number of pounds being converted by the number of dirham for £1.)

= 1230 AED

Example 8.2

Pietr lives in South Africa and has 500 ZAR which he is taking with him on holiday to China. How many Chinese yuan will he get for 500 ZAR?

1 ZAR = 0·51 CNY •———(Look down the left-hand column to find SA rand, then along to the Chinese yuan column. The '0·51' shows that for 1 rand you get 0·51 yuan.)

500 ZAR = 0·51 × 500

= 255 CNY

1 Jonathon has £250.

   a How many Swiss francs could he get for this?

   b How many euros could he get?

   c How many Chinese yuan could he get?

2 Alastair has 300 rand when he comes back from a South Africa holiday.

   He changes it back into GBP £.

   How many pounds would he get?

3 Mary sees a watch available from Qatar, in the Arab Emirates, costing 850 AED.

   She sees the same watch in Switzerland for 200 CHF.

   Which offers the better deal, and by how much?

4 George is going from Paris in Europe to Beijing in China.

   He takes €2000 with him.

   How many Chinese yuan will he get for his euros?

5 Alysoun is going on a two-city trip.

   She is going from her home in London to Berlin and then to Geneva in Switzerland, before returning home to London.

   She exchanges £750 to euros for her trip to Berlin.

   a How many euros does she get for her £750?

   b When she is there she spends €400.

      Alysoun then changes her remaining euros to Swiss francs for her visit to Geneva.

      How many euros does she exchange to Swiss francs?

   c How many CHF does she get for her euros?

   d In Geneva she spends 400 CHF.

      How many Swiss francs does she have left?

   e On her return she exchanges her Swiss francs back to British pounds.

      How many GBP does she get upon her return to London?

6 Johann is travelling from Cape Town in South Africa to New York in the USA and finally to London in Great Britain on a business trip.

   He leaves Cape Town with 30 000 ZAR.

   In New York he spends $850.

   He exchanges the remaining dollars for British pounds before going to London.

   How many GBP will he get for his dollars?

7 Diane is looking at some items on an internet shopping site.

She sees the same items available from different countries.

Which country offers the item for the cheaper price?

a Wooden statue: China for 400 CNY and Switzerland for 55 CHF

b Glass paper-weight: South Africa for 540 ZAR and France for 35 euros

c Light fitment: USA for $840 and Emirates for 3200 AED

> **Hint** Change one of each pair of currencies to the other currency so you can compare them. For example, in part **a** change 400 CNY to CHR or change 55 CHF to CNY.

8 Here are the currency exchange rates advertised by the Post Office on one particular day. The rates are slightly different if you exchange more money.

Spend	£400+	£500+	£1000+
Euro	1·0774	1·0801	1·0807
US dollar	1·2898	1·2931	1·2937
Australian dollar	1·5817	1·5883	1·5900
Swiss franc	1·2079	1·2142	1·2155
UAE dirham	4·6045	4·6190	4·6312

Source: Post Office Ltd

a How many euros would you get if you exchanged £550?

b How many US dollars would you get if you exchanged £1000?

c How many extra UAE dirham would you get if you exchanged £501 instead of £499?

9 Investigating the impact of interest rates on savings and borrowing

Exercise 9A Loans, savings, cards and credit agreements 🖩

N4 **1** The table shows the monthly repayments over 60 months and total cost of a loan of £10 000 from four different lenders. Use the table to answer the questions below.

Lender	Amount (£)	Interest rate (%)	Monthly payment (£)	Total repayment (£)
Hollifax	10 000	6·4	194·54	
Greene Loan	10 000	8·1	203·36	
Forthside Bank	10 000	10·4	215·52	
C2D Finance	10 000	12·7	222·51	

  a Copy this table and complete the total repayment column.

  Give your answers correct to the nearest penny.

  b How much more would you pay per month if you took a loan from C2D Finance rather than Hollifax?

  c Forthside Bank also offer a £10 000 loan over 25 years, where the repayment is £78 per month.

  How much extra would you pay for this loan compared to Forthside Bank's 60-month loan?

2 Jemima won £15 000 on a lottery draw.

She has two bank accounts she could put her winnings into.

- A Saver's account, which offers an annual interest rate of 3·4%

- An Ultra saver's account, which offers an annual interest rate of 5·1%

  a If she put the money into her Saver's account, how much interest would she receive after 1 year?

  b How much **more** interest would she get if she put her money into the Ultra saver's account for 1 year?

3 The Tait family want to borrow £3500 to go on holiday to Sydney, Australia.

They are comparing different loans from their bank.

	Annual interest charge	Loan term
The 1-year plan	8·7%	12 months
The $\frac{1}{2}$-year plan	7·8%	6 months

  a Calculate the monthly repayments for the 1-year plan.

  b Calculate the monthly repayments for the $\frac{1}{2}$-year plan.

  c Which loan should the Taits choose? Give a reason for your answer.

4 Marina spends £250 on her credit card. The card has a monthly interest rate of 3·2%.

a How much interest will be charged at the end of the first month?

b The equivalent annual percentage rate (APR) is 45·9%. How much interest would be due on the £250 if none had been paid for the year?

N5 **5** Aquarius Credit Card advertise an annual interest rate of 49%.

Calculate the interest to be paid at the end of 1 year on a loan of £3000.

6 Harry and Mhairi consider using a home improvement loan from a finance company to buy a new kitchen.

The finance company charge 27·5% simple interest on the loan amount.

Calculate the total amount to be repaid on a loan of £5000.

7 Ben has a balance of £632·50 on his credit card.

His minimum payment is £10 or 2·5% of the balance, whichever is greater.

a How much must Ben pay?

b After this is paid, 1·2% interest is added to the balance. How much does Ben now owe?

8 The Bulldog Building Society offers loans as advertised below.

Bulldog Building Society			
Small loan	£500–£4999	Interest rate	12·1%
Larger loan	£5000–£20 000	Interest rate	9·4%

Rebecca and Joe want to borrow £4900, over 1 year, to buy a new kitchen.

A friend says that it would be cheaper to borrow £5000.

Is the friend correct? Use your working to justify your answer.

9 Reuben invests £3100 in a savings account at a compound interest rate of 3·2% per annum.

How much interest will Reuben receive if he leaves the money in the account for 4 years?

10 Ying Lee is updating his bathroom. The work is priced at £2400.

He sees three adverts:

• Atlass Bank: a flat-rate loan of 10·1% repaid over 12 months

• Federalle Finance: a flat-rate loan of 6·8% repaid over 6 months

• Bathroom Store Card: a 5% discount on price, then interest of 12·7% charged on the balance repaid over 12 months.

a Which has the lowest monthly payment?

b Which has the lowest total payment?

11 Justin bought a house in 2013 for £121 000.

For the first 3 years the value of the house fell by 3% per annum.

The next 2 years it rose in value by 5% per annum.

What is Justin's house worth in 2018? Give your answer correct to the nearest £100.

10 Using a combination of statistics to investigate risk and its impact on life

Exercise 10A Using the link between simple probability and expected frequency

N3 **1** In a lottery draw, there are 49 balls numbered from 1 to 49.

 a What is the probability of drawing the ball numbered 9?

 b What is the probability of drawing a ball which has a 5 on it?

 c What is the probability of picking a number which is a multiple of 10?

2 John rolls a standard 1–6 dice.

 a What is the probability of rolling a 3? Write your answer as a decimal fraction.

 b What is the probability of rolling an even number? Write your answer as a percentage.

3 Arianne will win if the spinner lands on a 3.

 What are Arianne's chances of winning?

 Copy this probability scale and mark your answer on it.

```
├────────────┬────────────┤
0          0·5           1
```

N4 **4** A primary teacher does a survey in her class on how pupils travelled to school and if they were wearing uniform. The results are shown in the table.

	Uniform	No uniform
Walk	15	2
Bus	10	3

 a What is the probability that a pupil, picked at random, walked to school and was not wearing uniform?

 b What is the probability that a pupil, picked at random, was wearing uniform and came by bus?

5 There are 5 red balls, 7 green balls, 4 yellow balls and 2 white balls in a bag.

 a What is the probability of randomly picking a green ball?

 b What is the probability of randomly picking either a red or a yellow ball?

 c Cory dipped into the bag 36 times, noted the colour of the ball he picked and returned it to the bag.

 How many white balls would he expect to pull out?

6 The weather forecast says that the chance of rain on any day in April is 2 in 5.

 On how many days in April would you expect rain to fall?

7 In a charity raffle, 300 tickets were sold. Jennifer bought 10 tickets.

 a What is the probability that Jennifer will win a prize?

 b There are 5 prizes. What is the chance that Jennifer has at least one winning ticket?

8 The traffic police did a survey on cars going around a roundabout.

The results are shown in the table below.

	Used indicator	Did not use indicator
In correct lane	70	12
Not in correct lane	8	10

a What is the probability that a car was in the correct lane and the indicator was used?

b Drivers who did not use indicators and were in the wrong lane were stopped.

Using the survey data, how many cars would you expect the police to stop if they observed 200 cars?

N5 **9** Ms Edmonds did a survey with her Higher Maths class on which social media they used most. The results are as shown in the table.

	Facebook	Instagram	WhatsApp
Boys	9	3	2
Girls	4	11	1

What is the probability that a pupil, picked at random, is a boy who uses Instagram the most? Give your answer as a fraction in its simplest form.

10 To win a game, Jeremy needs the spinners to land on either two 1s or two 6s.

What is the probability that Jeremy will win the game? Give your answer as a fraction in its simplest form.

11 In the 2017 Athletic World Championships, France and China won 12 medals between them.

China won 3 silver medals, and an equal number of gold and bronze medals.

France won 5 medals in total, none of them silver.

France won one more gold than bronze.

a Copy and complete the table to show the medal results.

Country	Gold	Silver	Bronze	Total
France				
China				
Total				12

b If China picks one of its 2017 medal winners at random, what is the probability it will be a bronze medal winner?

12 Harry's Quidditch Club sells 400 raffle tickets. Harry buys 12 tickets.

Harry's Owl Club sells 350 raffle tickets. Harry buys 11 tickets.

In which raffle is Harry more likely to win a prize? Use your working to justify your answer.

11 Using a combination of statistical information presented in different diagrams

Exercise 11A Constructing, interpreting and comparing different representations of data

   1 A survey was done outside Charley's Electrical Store, asking people which TV channel they watched the most. The results are shown in the table.

Channel	BBC1	BBC2	ITV1	ITV4	SKY1	Sky Sports	MTV	Scyfy
Number of viewers	12	8	10	3	11	13	7	9

Draw and label a bar graph to show this information.

> **Hint** Remember when drawing graphs you should:
> - have a title
> - label the axes
> - choose suitable scales
> - draw neatly and accurately.

 2 Maria kept track of the number of goals she scored whilst playing football, over a six-month period.

Her results are shown in the table.

Month	Number of goals
November	10
December	8
January	7
February	12
March	14
April	5

Copy and complete this pictograph to show Maria's goals.

	Number of goals
November	
December	
January	
February	
March	
April	

⚽ = 2 goals

3 As part of research into healthy eating, school pupils were asked which was their favourite fruit. The results are shown in the table.

Fruit	Number of pupils
apple	10
pear	5
orange	8
peach	1
banana	13

Draw a neat, labelled, bar graph to show this information.

 4 A school recorded the number of absences in S1 in the week leading up to the Christmas holidays. The results are shown in the table.

Day	Monday	Tuesday	Wednesday	Thursday	Friday
Number of pupils absent	15	22	13	31	42

Draw a neat, labelled, line graph to display this information.

 5 Ladl supermarket has some vegetables on display.

Copy and complete the tally table for this display.

Vegetable	Tally mark	Frequency
🍅		
🫑		
🌶		

N4 **6** The Broad family have a monthly income of £3000.

They have drawn a pie chart to show how the money is spent.

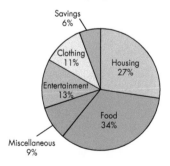

a How much do the Broads spend on housing?

b How much do they spend on clothing?

c What would be the Broad's monthly spend on food?

7 Gemma is an amateur weather watcher.

She recorded the average monthly temperature over 1 year.

The results are shown in the graph.

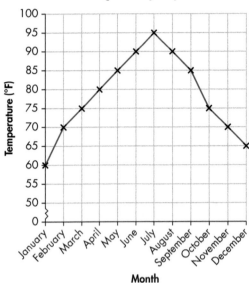

Use the graph to answer these questions.

a What was the highest average temperature recorded?

b When was there a fall of 10 °F?

c Which was the coldest month?

d Which months recorded a temperature of 85 °F?

8 A sample of 20 cars was taken from a car park, and the colour of each car was noted.

The results were as follows:

red   white   silver   blue   silver   silver   yellow   red   white   silver

silver blue   red   red   blue   white   silver   red   blue   white

a Copy this table and complete the tally and frequency columns.

Colour	Tally	Frequency
red		
blue		
silver		
white		
yellow		

b Draw a neat, labelled, bar chart to show this information.

9 Reginald was growing a sunflower.

Each week, for 6 weeks, he measured its height. The table shows the results.

Week	Week 1	Week 2	Week 3	Week 4	Week 5	Week 6
Height (cm)	20 cm	30 cm	50 cm	60 cm	65 cm	70 cm

Construct a line graph to illustrate this information.

10 The percentage of types of cars sold by a car showroom were:

Family saloons   30%

Hatchbacks   25%

4 × 4s   40%

Sports   5%

Construct a pie chart to illustrate this information.

Example 11.1

Here are the marks scored by 15 pupils in a maths test.

78 82 74 45 68 75 93 54 61 70 48 66 62 51 77

a Write down the median and lower and upper quartiles.

b Draw a boxplot to illustrate the maths marks.

a 45 48 51 (54) 61 62 66 (68) 70 74 75 (77) 78 82 93 •——— Rewrite the numbers in numerical order.

Median, Q2 = 68 •——— The median (Q2) is the middle number.

Lower quartile, Q1 = 54 •——— The lower quartile (Q1) is the median of the lower half of the data set.

Upper quartile, Q3 = 77 •——— The upper quartile (Q3) is the median of the upper half of the data set.

b

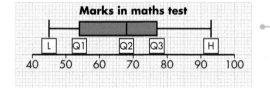

Use the median, lower quartile, upper quartile, and lowest and highest values to draw a boxplot.

  11 Miss Bryant measured the heights of a group of S2 girls. The heights, in centimetres, are:

153   147   160   146   162   158   159   149   152   150   163

a Draw a boxplot to show the heights of the girls.

A boxplot for a group of S2 boys is shown.

b Make two valid comments comparing the heights of the boys and the girls.

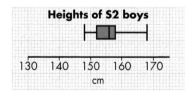

12 A class sat a maths test and a physics test. The marks are shown in the table.

Pupil	A	B	C	D	E	F	G	H
Maths mark (%)	56	73	51	61	24	71	94	34
Physics mark (%)	62	68	44	67	36	73	87	33

a Plot these points on a scatter graph.

b Draw a best-fitting line.

c Is there a correlation between the marks in maths and physics?

d Pupil I scores 83 marks in his maths test but was absent for the physics test.

Use your best-fitting line to estimate his score in the physics test.

12 Using statistics to analyse and compare data sets

Exercise 12A Calculating different types of average, range and standard deviation

> **Hint** Remember how to calculate three different averages.
>
> $$\text{Mean} = \frac{\text{total}}{\text{number of items of data}}$$
>
> Median = middle number, when data is put in ascending order
>
> Mode = most common or frequent item of data

N3 **1** The lengths of 10 stick insects, in centimetres, were recorded as:

   6   5   5   7   8   3   4   4   6   7

   a Calculate the total length of the stick insects.

   b Calculate the mean length of the stick insects.

N4 **2** Mandy measured the temperature, in °C, each day whilst she was on holiday.

   The temperatures recorded were 38, 44, 41, 32 and 35.

   a What was the range of the temperatures?

   b What was the mean temperature?

3 Sanjid weighed 5 packets of icing sugar.

   The weights were 46 g, 43 g, 37 g, 49 g and 40 g.

   Calculate the mean weight.

4 Tanya weighed 7 sacks of potatoes.

   The weights, in kilograms, were recorded as:

   11·2   14·3   7·8   0·850   6·3   0·670   0·420

   Calculate the mean weight of the sacks of potatoes.

5 Rory's batting scores in a cricket match were recorded as:

   121   85   15   88   62   71

   a What was the range of Rory's scores?

   b What was his mean score? Give your answer correct to 2 decimal places.

6 Arlene sells cars. She will get a bonus if the mean value of cars sold in a week is at least £6000.

   One week Arlene sold cars to the value of £7000, £4570, £5800, £9999, £3900, £5990 and £6250.

   Will Arlene get a bonus for this week?

   Use your working to justify your answer.

Example 12.1

Said weighed 4 boxes of chocolates.

The weights, in grams, were: 450, 400, 425 and 375.

a Calculate the mean weight of the boxes of chocolates.

b Calculate the standard deviation for the boxes of chocolates.

a Mean $= \dfrac{450 + 400 + 425 + 375}{4}$

$= \dfrac{1650}{4}$

$= 412 \cdot 5 \, \text{g}$

b Standard deviation

$s = \sqrt{\dfrac{\sum (x - \bar{x})^2}{n - 1}}$

> The standard deviation is the square root of the average of the squares of differences between each of the numbers, x, and the mean of the numbers, $\bar{x}$.

x	$\bar{x}$	$x - \bar{x}$	$(x - \bar{x})^2$
450	412·5	37·5	1406·25
400	412·5	−12·5	156·25
425	412·5	12·5	156·25
375	412·5	−37·5	1406·25
Total			3125

$s = \sqrt{\dfrac{3125}{3}}$

> Substitute your values into the formula.

$= \sqrt{1041 \cdot 67}$

$= 32 \cdot 27 \, \text{g}$

7 Tony weighs himself each day for a week.

The results, in kilograms, are recorded as follows:

87·1, 87·9, 86·9, 87·0, 87·5, 87·2, 87·5

a What is Tony's mean weight?

b Calculate the standard deviation.

8 Danni is a holiday salesperson.

She wants to advertise a holiday and use the best measure of average to persuade people to buy a holiday. She measures the hours of sunshine per day, over a week, as:

10   8   9   6   6   11   7

a Calculate:

   i   the mean

   ii the median

   iii the mode.

b Which of these should she use in the advert?

9 The top ten all-time men's 100 m sprint times in athletics are shown in the table.

	Athlete	Where	Year	Time for 100 m (in seconds)
1	Usain Bolt	Berlin	2009	9·58
2	Usain Bolt	London	2012	9·63
3=	Usain Bolt	Beijing	2008	9·69
3=	Tyson Gay	Shanghai	2009	9·69
3=	Yohan Blake	Lausanne	2012	9·69
6	Tyson Gay	Berlin	2009	9·71
7=	Usain Bolt	New York City	2008	9·72
7=	Asafa Powell	Lausanne	2008	9·72
9=	Asafa Powell	Rieti	2007	9·74
9=	Justin Gatlin	Ad-dawah	2015	9·74

Calculate the mean and standard deviation for these times to see how consistent the times have been over the last 10 years.

10 Mary records the number of calories in a serving of different cereals.

Her results are: 135, 115, 120, 110, 110, 100, 105, 110, 125.

Calculate the mean and standard deviation for the calories in these cereals.

11 The waiting times, in minutes, of 6 people waiting at a garage service centre are recorded below.

   20   25   18   19   16   22

a Calculate the mean waiting time.

b Calculate the standard deviation.

c For another group of people, at a nearby service centre, the mean waiting time was 16·25 minutes and the standard deviation was 2·40 minutes.

   Make two valid comparisons between the waiting times at these service centres.

12 Kylie is a quality control inspector in a juice bottling plant.

One day she takes a sample of 6 bottles of juice and measures the volume of juice, in millilitres, in each bottle.

Her results are: 401, 394, 407, 391, 405, 402.

a Calculate the mean and standard deviation for the volumes of juice.

b Kylie adjusts the machinery in an attempt to make the bottling more consistent.

   After taking another sample, Kylie finds the mean to be 400 ml and the standard deviation to be 5·8 ml.

   Has her adjustment been successful?

   Use your working to justify your answer.

13 Stefan grows tomatoes in an industrial greenhouse.

He installs a heating system to maintain the temperature.

Stefan will regard the system as efficient if the mean temperature is $20 \pm 0.6\,°C$ and the standard deviation is less than 2 °C.

Stefan records the temperature over a period of days to check the efficiency of his heating system. His results, in degrees Celsius, are: 20, 21, 19, 21, 23, 19.

Is Stefan's heating system working efficiently?

Use your working to justify your answer.

13 Drawing a best-fitting line from given data

Exercise 13A Data presented in tabular form

 1 The table shows the marks scored by pupils in an English and a history test.

English test	25	35	30	78	45	67	91	82	60
History test	40	37	36	75	60	85	87	93	55

   a Draw a scatter graph to show this data.

   b Draw a best-fitting line for the data points.

   c If a pupil scored 52 marks in English, what would you expect them to score in the history test?

   > **Hint** When drawing a best-fitting line, think about the slope and try to get about the same number of points above and below the line.

2 The table shows the number of hours of sunshine and the amount of rainfall, in millimetres, at a holiday resort.

Hours of sunshine	6	9	6	7	4	5	8	7	8
Rainfall (mm)	0	1	5	2	10	10	2	3	1

   a Draw a scatter graph to show this data.

   b Draw a best-fitting line for the data points.

   c One day there were 6·5 hours of sunshine.

      How many millimetres of rain would you expect on that day?

3 The table shows the resting heartbeat in beats per minute (bpm) and the number of hours per week that a group of people go to the gym.

Heartbeat (bpm)	75	87	55	60	60	81	92	70
Time in gym (hours)	7	3	12	10	12	6	4	8

   a Draw a scatter graph to show this data.

   b Draw a best-fitting line for your points.

   c Candice goes to the gym for 9 hours per week.

      What would you expect Candice's resting heartbeat to be?

Mr Sands did a survey with 10 pupils in his N5 Maths class.

He asked them how many hours of revision they did in the week before they sat a prelim exam. He also recorded the mark each pupil got for their prelim.

The results are shown in the table.

Pupil	1	2	3	4	5	6	7	8	9	10
Revision time (hours)	3	2	4	6	5	2	8	3	2	1
Prelim mark	75	65	70	85	65	55	80	50	60	30

a Draw a scatter graph to show this data.

b Draw a best-fitting line for the data points.

c Pupil 8 would like to improve her mark to 65% for the next test.

How many **extra** hours should she revise before her next test if she is to achieve this?

5 A group of friends are doing a cycle ride to raise money for charity.

The distance each friend cycled, and the amount of money raised, is shown in the table.

Name	Distance cycled (km)	Amount raised (£)
Jack	30	120
Jill	50	200
Frances	20	40
Joseph	40	100
Aretha	30	80
Franklin	25	150
Adam	5	90

a Draw a scatter graph to show this data.

b Draw a best-fitting line for the data points.

c Davina aims to cycle 35 km.

How much money would you expect her to raise for charity?

14 Working with simple patterns and calculating a quantity

Exercise 14A Working with simple patterns

 1 Gavin is laying hexagonal paving slabs in his garden.

Make a template of the slab, shown on the right, or ask your teacher for one.

Show how Gavin could cover the area in his garden, shown below.

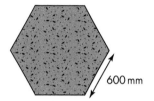

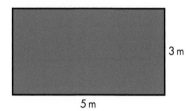

2 Draw the next two designs in these patterns.

a

b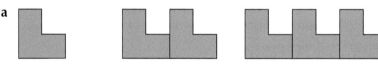

c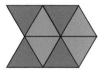

3 Jenny is designing a pattern for wallpaper.

Using this shape as a template, show how Jenny could continue this pattern on a roll of wallpaper.

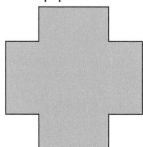

Exercise 14B Calculating a quantity based on two related pieces of information 🖩

N4 **Example 14.1**

8 km is approximately equal to 5 miles. What is 50 km equal to in miles?

km		miles
8	→	5
1	→	$5 ÷ 8 = 0·625$ miles
50	→	$0·625 × 50 = 31·25$ miles

Find the equivalent of 1 km in miles.

Multiply by 50 to find the equivalent of 50 km in miles.

N4 **1** Graeme is 6 ft tall, which is equivalent to 1·83 m. A cricket wicket is 66 ft long.

What is the length of a cricket wicket in metres?

2 James bought a 25 lb bag of potatoes. The label showed 25 lb = 11·34 kg.

What would be the weight, in kilograms, of a bag of onions weighing 2 lb?

3 Janice is using an old recipe for a carrot cake. The recipe asks for 9 ounces of grated carrot.

How many grams of grated carrot is needed if 1 ounce is approximately 28·35 grams?

4 Khan changed £50 to euros whilst on a cruise liner. For £50 he got €56·47.

His breakfast on the liner came to £12·50.

If Khan paid in euros, how much would he have to pay?

N5 **5** Jack put 20 l of fuel in his car. The printout showed that 20 l = 4·40 gal.

The following week Jack put 65 l of fuel in his car.

How many gallons would that be?

6 Jenna is putting air into her car tyre. The gauge gives readings in pounds per square inch, psi, and also in bars.

Jenna pumps her tyre up to 33 psi, which shows as equal to 2·28 bar.

Jenna has a racing bicycle and wants to pump her tyres up to 70 psi.

The pump she is using only shows measurement in bars.

What should she pump her tyre to?

7 Gena notices that her utility bill says:

Energy used: 10 kWh = 0·34 therm

How many therms would be used if 8 kWh were used?

> **Hint** kWh is a kilowatt hour, which is equivalent to 1000 watts applied for 1 hour.

8 On a map, $\frac{3}{8}$ of an inch represents 60 kilometres.

How many kilometres would 2 inches represent?

9 Boats measure their speed in knots. 1 knot is approximately 1·15 mph.

The maximum speed for a canal boat is 4 mph.

What is this speed in knots? Give your answer correct to 2 decimal places.

15 Constructing a scale drawing, including choosing a scale

Exercise 15A From written information and/or a sketch

N3 **1** Jennifer has sketched a kite on a squared grid.
Draw the kite, enlarged by a scale factor of 2.

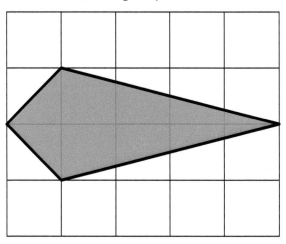

2 This sketch of a tree is drawn to a scale of 1 cm to 2 m.
What is the real height of the tree?

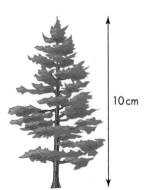

10 cm

3 Mark is making a model of a space shuttle as shown below.
The scale of the model is 1 cm to 7 m.
a What is the real length of the shuttle?
b What is the real wingspan of the shuttle?

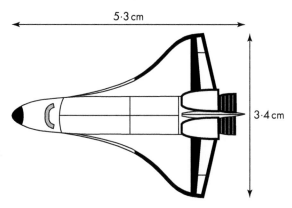

5·3 cm

3·4 cm

4 This is a street map of the centre of Newtown.

The scale is 1 cm to 25 m.

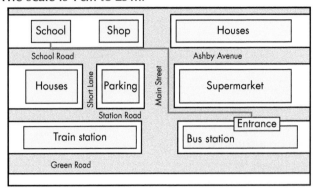

Beatrice comes out of the school, walks along School Road, turns into Main Street, walks down and then turns left into Station Road to the bus station entrance.

The route is shown in red on the map.

Estimate how many metres Beatrice has walked.

N4 **5** Abbie is making a nest of tables in her Craft and Design class.

The set of three tables have to fit into each other.

The smallest table has these dimensions:

length 400 mm   breadth 300 mm   height 350 mm

The scale factor from one table to the next is 1·2.

Calculate the length, breadth and height of the other two tables.

6 Darrett, the house builders, have bought a plot of land for building some houses.

Their surveyor has made a rough sketch of the plot.

Using a scale of 1 cm to 50 m, draw an accurate plan of the plot.

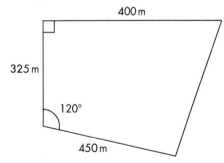

7 Craig has made a scale drawing of the Eiffel Tower.

His sketch is 13·5 cm tall.

The real Eiffel Tower is 324 m tall.

What scale did Craig use?

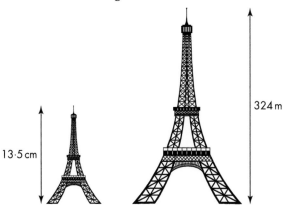

13·5 cm

324 m

8 A ship leaves port and sails on a bearing of 050° for 100 km.

It then turns and heads on a new bearing of 210° for 200 km.

a Using a scale of 1 cm:20 km, construct a scale drawing to illustrate this journey.

b The ship then heads back to port.
  On what bearing should the ship set? How far will it travel?

9 A marine research aeroplane is following a school of whales in the ocean.

The diagram shows the route the aeroplane followed.

Starting from island *A* it flies on a bearing of 105° for 25 km to island *B*.

From island *B* it flies on a bearing of 075° for 30 km to island *C*.

The aeroplane then flies directly back to island *A*.

Using a suitable scale, construct a scale drawing and calculate the total distance covered by the aeroplane.

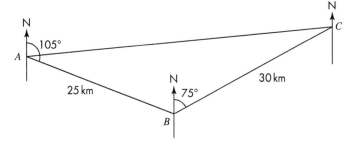

10 Kara is making a sail for her model boat.

She makes a rough sketch of the sail.

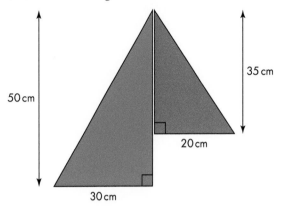

a Using a suitable scale, make an accurate scale drawing of the sail.

b Calculate the area of material required to make the sail for the model.

11 The bearing of oil platform Beta from ship Alpha is 080°. They are 50 km apart.

a Using a suitable scale, construct a scale drawing to illustrate this.

b A customs vessel Gamma is on a bearing of 130° from ship Alpha and 215° from oil platform Beta. Mark its position on your diagram.

c How far is Gamma from oil platform Beta?

16 Planning a navigation course

Exercise 16A From written information and/or a sketch

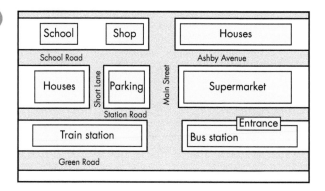

Mark is giving his friend, Alicia, directions on how to get from the bus station entrance to the school.

He says 'come out of the bus station, turn left. Go along Station Road, then turn right into Short Lane. At the end of Short Lane, turn left. The school is on your right.'

a Copy the diagram and mark in the route which Alicia will take.

b Write down another set of directions which would get Alicia from the bus station to the school.

 An oil platform, Omega, is 40 km from port Phi on a bearing of 140°.

A platform support vessel, Vega, is 35 km from Omega on a bearing of 250°, as shown in the sketch.

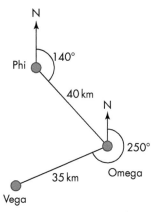

a Make an accurate scale drawing using a scale of 1 cm to 5 km.

b How far is the platform support vessel from the port?

c If the support vessel wanted to go to port, on what bearing should it head?

3 The map shows part of the area around the Isle of Wight.

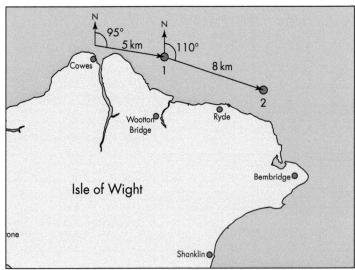

A sailing boat leaves Cowes and sails on a bearing of 095° for 5 km to buoy 1.

It then turns and heads towards buoy 2, 8 km away on a bearing of 110°.

a Make an accurate scale drawing using a scale of 1 cm to 1 km.

b The sailing boat wants to return directly to Cowes.

On what bearing and for what distance does it need to sail?

4 Jenny is in the school orienteering team and is going on a practice run.

The first part of her run is shown in the diagram.

Scale: 1 cm to 50 m

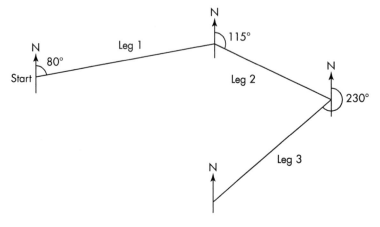

Copy and complete this table for Jenny's run.

	Bearing	Distance (metres)
Leg 1		
Leg 2		
Leg 3		
Back to start		

5 This is a map of Western Australia.

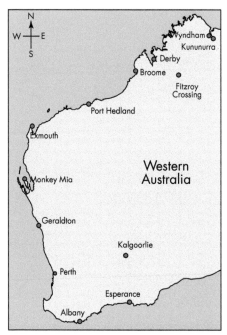

An Outback trucker is carrying a load from Perth.

He drives to Port Hedland, on to Fitzroy Crossing and then to Kalgoorlie before returning to Perth.

To get an idea of the length of his journey, the trucker maps out direct distances and bearings.

The distances and bearings are shown in the table.

Journey	Bearing	Distance (km)
Perth to Port Hedland	012°	1320
Port Hedland to Fitzroy Crossing	073°	770
Fitzroy Crossing to Kalgoorlie	197°	1455
Kalgoorlie to Perth	255°	550

Using a scale of 1:40 000 000, make an accurate scale drawing of his journey.

6 A speedboat leaves a port on a bearing of 060° and travels at a speed of 80 km/h for 45 minutes.

It then turns on a new bearing of 210° and travels at the same speed for 1 hour 24 minutes.

a Using a scale of 1 cm:20 km, construct a scale drawing to illustrate this journey.

b The speedboat continues at the same speed back to port.

Use your scale drawing to determine the distance and bearing of the port from the speedboat.

7 A Navy search and rescue helicopter was out on patrol.

The diagram shows the route the helicopter followed.

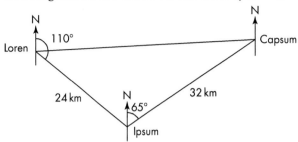

From Loren it flew on a bearing of 110° for 24 km to Ipsum.

From Ipsum it flew on a bearing of 065° for 32 km to Capsum.

The helicopter then flew directly back to Loren.

a Using a suitable scale, construct a scale drawing and calculate the total distance travelled by the helicopter.

b The helicopter flew at an average speed of 60 km/h.

It carried enough fuel for 4 hours' flying time.

What percentage of fuel did the helicopter use during this flight?

8 The map shows the ports of Stockholm in Sweden, Tallinn in Estonia, and Helsinki in Finland.

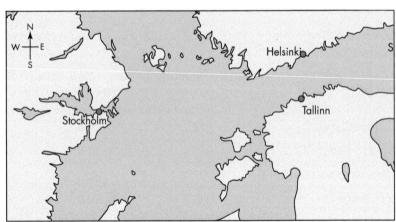

From	To	Bearing	Distance
Stockholm	Helsinki	077°	400 km
Stockholm	Tallinn	089°	380 km

a Use a suitable scale to draw a scale map of these three ports.

b What is the bearing of Helsinki from Tallinn?

c A fishing ship is on a bearing of 300° and a distance of 200 km from Tallinn.

Mark the position of the ship on your scale drawing.

d What is the bearing and distance of Stockholm from the fishing ship?

9 The map shows four airports in Scotland.

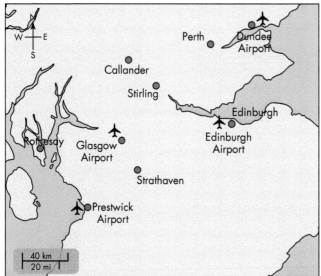

a What is the bearing and distance of Dundee Airport from Glasgow Airport?

b What is the bearing and distance of Edinburgh Airport from Prestwick Airport?

c A small private aeroplane is flying from Dundee Airport to Prestwick Airport, via Glasgow.

 If its average speed is 170 km/h, how long will the flight take?

17 Carrying out efficient container packing

Exercise 17A By assigning items to uniform containers to minimise amount of containers used

N4 | **Example 17.1**

Jason is cutting lengths of fabric for scenery in a stage play.

Fabric comes in 10 m lengths.

The lengths Jason requires are shown in the table.

Fabric	A	B	C	D	E	F	G	H	I	J
Length (m)	7	8	4	4	2	2	3	5	8	3

Use the first-fit algorithm to determine how many 10 m lengths Jason will need.

Length 1	Length 2	Length 3	Length 4	Length 5	Length 6
A	B	C	G	I	J
			H		
		D			
E					
	F				

> First-fit means you put the item, as you meet it, in the first available space.

So Jason will need six 10 m lengths.

Example 17.2

Jason thinks he can be more efficient with his cuts.

Using a decreasing first-fit algorithm, can you reduce the number of lengths of fabric required?

Fabric	A	B	C	D	E	F	G	H	I	J
Length (m)	7	8	4	4	2	2	3	5	8	3

Fabric	B	I	A	H	C	D	G	J	E	F
Length (m)	8	8	7	5	4	4	3	3	2	2

First, re-order the fabric lengths in order of size, starting with the largest.

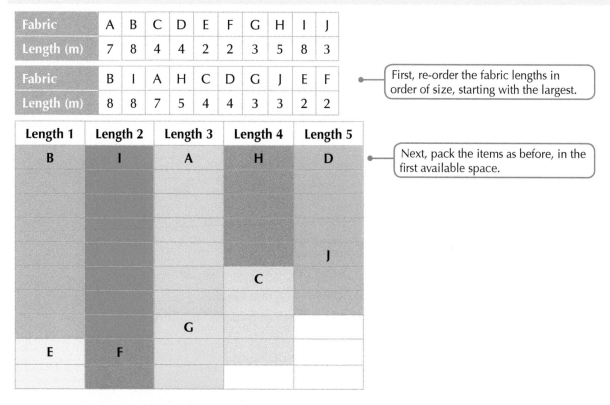

Next, pack the items as before, in the first available space.

Jason will only need five lengths of 10 m.

Use the first-fit algorithm to answer Questions 1–5.

1 James is moving from home to a flat. He has 8 items to pack.

He has packing boxes which will hold 8 kg each.

The weights of his items are shown in the table.

How many boxes will James require?

Hint Draw a few columns, as in Example 7.1, and label them Box 1, Box 2, Box 3, etc. Then start to 'pack' the items, as you meet them, into the first available space.

Item	A	B	C	D	E	F	G	H
Weight (kg)	1	2	3	1	3	4	2	1

2 The table below shows the lengths, in minutes, of soundtracks from films.

They are to be recorded onto tapes which each have a recording time of 100 minutes.

How many tapes will be required to record all the soundtracks?

Track	1	2	3	4	5	6	7	8	9
Length (min)	29	52	73	87	74	47	38	61	41

3 The table below shows the weights, in kilograms, of factory machines.

They need to be transported to the factory in crates.

Each crate can hold a maximum weight of 60 kilograms.

How many crates will be required to transport all the machines?

Machine	M1	M2	M3	M4	M5	M6	M7
Weight (kg)	41	28	42	31	36	32	29

4 Jake is working on fabric for a big stage production.

Fabric comes in rolls of length 60 metres.

The lengths Jake needs are listed below.

How many rolls of fabric will Jake require?

Fabric	F1	F2	F3	F4	F5	F6	F7	F8	F9	F10
Length (m)	32	45	17	23	38	28	16	9	10	12

5 Karen, a carpenter, is making some bedside furniture.

She needs 9 pieces of wood, as listed below, to make the furniture.

The wood shop sells planks of wood in 1000 mm lengths.

How many planks will Karen need to make her bedside furniture?

Wood	W1	W2	W3	W4	W5	W6	W7	W8	W9
Length (mm)	200	200	200	350	400	500	600	700	750

N5 **6** For Questions 1–5, use a decreasing first-fit algorithm to show and describe whether this is a more efficient method than the first-fit algorithm.

7 Eddie is a lorry driver, delivering computer cabinets.

The cabinets have dimensions as shown and must be loaded upright.

The lorry has two shelves with dimensions as shown.

Calculate the maximum number of computer cabinets which can be loaded on to the lorry.

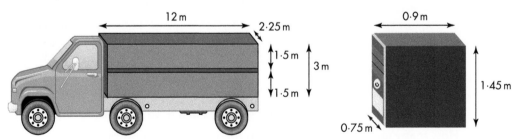

18 Using precedence tables to plan tasks

Exercise 18A Constructing and using precedence tables

5 | **Example 18.1**

Arabella and Andrew are building a garden shed.

They plan out what they need to do and set it out in a precedence table, as shown.

Activity	Description	Preceded by	Time (hours)
A	Prepare foundations	–	3
B	Position door frame	A	1
C	Erect walls	A, B	4
D	Fit electric power	C	2
E	Put on roof	D	2
F	Fit windows	C	2
G	Fit gutters	E	1
H	Paint shed	G	4

a Draw a network diagram to illustrate this.

b What would be the minimum time to complete the job?

a

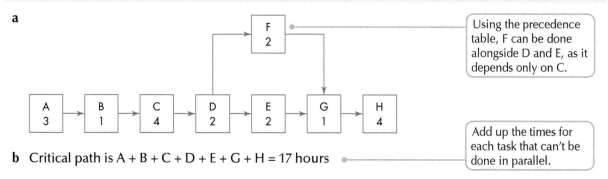

Using the precedence table, F can be done alongside D and E, as it depends only on C.

Add up the times for each task that can't be done in parallel.

b Critical path is A + B + C + D + E + G + H = 17 hours

An advertising campaign is planned to promote a new perfume made by the company Elvin.

A precedence table has been created to plan the launch of the campaign.

Task	Task detail	Preceding task(s)	Time (days)
A	Plan the campaign	–	4
B	Make an advert for television	A	6
C	Design a poster for subways	A	7
D	Carry out market research on TV advert	B	8
E	Carry out market research on poster	C	10
F	Management to review market research	D, E	9
G	Campaign goes to marketing staff to prepare launch	D, E, F	5

a Prepare a network diagram to illustrate this information.

b If Elvin want to launch the new perfume on Valentine's Day, 14 February, when should the company begin planning the campaign?

2 Mike is making a chilli con carne for the evening meal.

The table shows the tasks to be done and the time taken for each task.

Task	Task detail	Preceding task(s)	Time (minutes)
A	Prepare vegetables and kidney beans	–	10
B	Lightly cook vegetables in a pot	A	5
C	Add mince to pot to brown	B	12
D	Add tomatoes and kidney beans and simmer	C	35
E	Boil a pot of water	A	5
F	Add rice to pot of water and cook	E	12
G	Put rice and chilli onto plates	D, F	5
H	Sprinkle some parsley leaves on top	G	2
I	Serve meal	G	4

a Construct a network diagram to illustrate this.

b Write down a critical path to show the minimum time it would take Mike to prepare the meal.

c If Mike wants to serve the meal at 8.30 pm, when should he start to prepare?

3 Emma is making breakfast.

She lists the tasks she needs to do, and the time for each.

warm plates (2 min)

grill bacon (4 min)

make toast in toaster (3 min)

fry eggs (3 min)

warm up grill (2 min)

heat oil in frying pan (1 min)

buy bacon and eggs (15 min)

serve (5 min)

a Construct a precedence table, putting the tasks in a reasonable order and clearly show where a task must be preceded by another.

b If Emma wants to serve breakfast at 0730, what is the latest time she should start?

4 The Digby family are employing Murdo's bathroom specialists to redesign their bathroom. Murdo's provide a team of workers to refit the bathroom. The table shows the list of tasks and the time taken for each.

Task	Task detail	Preceded by	Time (hours)
A	Prepare plumbing	–	4
B	Prepare electrics	–	4
C	Construct vanity units	–	2
D	Plaster walls	A, B, C	6
E	Fit basin, bath, shower	D	3
F	Fit shower cabinet	D	2
G	Tile walls	F	8
H	Fit flooring	G	5
I	Complete plumbing	H	3
J	Complete electrics	H	2
K	Tidy up	I, J	1

a Copy and complete the diagram below to show the tasks and times taken.

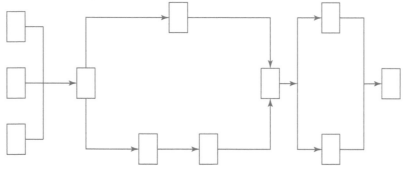

b Murdo's tell the Digbys they will fit the bathroom in 27 hours.

Is this a valid claim?

Give a reason for your answer.

19 Solving a problem involving time management

Exercise 19A Planning the timing of activities with complex features

N3 **Example 19.1**

The Smith family left their home in their car at 0915 to travel to their holiday hotel.

They arrived at the hotel at 1750.

How long did their journey take?

0915   1000   1700   1750

   45 min   7 hours   50 min

> Draw a timeline, marking major times. Work out the interval between each time.

Total time = 7 hours 95 minutes

> 95 minutes = 1 hour 35 minutes

        = 8 hours and 35 minutes

1 This is part of the train timetable from Edinburgh to London.

How long does each train take to travel from Edinburgh to London Kings Cross?

Edinburgh to London Kings Cross	
Departs	**Arrives**
15:30	19:50
16:00	20:46
16:30	20:52

2 The timetable shows some of the buses from Edinburgh to Glasgow.

Edinburgh	Glasgow				
Edinburgh bus station	0515	0545	0615	0630	0645
Edinburgh Haymarket Station	0525	0555	0625	0640	0657
Corstorphine opp Zoo Park	0530	0601	0631	0646	0703
Edinburgh Airport	▼	▼	▼	▼	▼
Ratho Station Road End	0541	0612	0642	0657	0714
Eurocentral Maxim Business Park	0614	▼	0715	▼	0747
Baillieston	▼	▼	▼	▼	▼
Glasgow Buchanan bus station	0634	0706	0740	0756	0815

Timothy wants to be in Glasgow, Buchanan bus station, for 8 am.

What is the latest time he can catch a bus from Edinburgh bus station?

3 A television programme started at 1615 and finished at 1750.

How long did the programme last?

4 A train leaves London Paddington station at 7.55 am.

It arrives in Edinburgh Waverley at 1.12 pm.

Calculate the time taken by the train for this journey.

N4 **5** An overnight coach leaves Perth at 2015 and arrives in London at 0810.

How long does the journey take?

6 **a** On 3 October 2017, sunrise in Fort William was at 0728.

The sun set 11 hours and 24 minutes later.

At what time did the sun set?

b On the same day in Fort William, the moonrise was at 1803.

The moon set the following day at 0404.

For how long was the moon visible?

7 Iceland is in the same time zone as the UK.

Craig is flying from Edinburgh to Reykjavik, Iceland.

He leaves Edinburgh at 2245 and the flight takes 3 hours and 20 minutes.

At what time does Craig land in Reykjavik?

> Hint | If two places are in the same time zone then clocks show the same time: 3 pm in the UK is 3 pm in Iceland.

8 The time in Athens is 2 hours ahead of the time in Dundee.

This means that 1200 in Inverness is 1400 in Athens.

Jasmine leaves on a small plane from Dundee at 1130 to fly to Athens.

The flight takes 4 hours 30 minutes.

What is the local time when Jasmine lands at Athens?

N5 **9** The map shows part of Europe and the time zones across it.

Western Europe	GMT
Central Europe	GMT + 1
Eastern Europe	GMT + 2

Donald is in London.

He wants to phone Vladimir in Poland, Eastern Europe, and Emmanuel in France, Central Europe, on a conference call.

Donald's clock says 1430.

What times show on Vladimir's and Emmanuel's clocks?

10 Brooklyn arrived at Glasgow airport for her 1315 flight to Johannesburg.

The signs indicated that the flight was delayed by 1 hour 15 minutes.

Johannesburg is 2 hours ahead of GMT.

The flight time is 14 hours and 25 minutes.

At what (local) time will Brooklyn arrive in Johannesburg?

Use the world time zones table shown below to answer Questions 11–15.

City, Country	GMT ±	City, Country	GMT ±
Las Vegas, USA	– 8 hours	Moscow, Russia	+ 3 hours
Denver, USA	– 7 hours	Lahore, Pakistan	+ 5 hours
Chicago, USA	– 6 hours	New Delhi, India	+ 5 hours 30 min
New York, USA	– 5 hours	Beijing, China	+ 8 hours
Mexico City, Mexico	– 2 hours	Hong Kong	+ 8 hours
Rio de Janeiro, Brazil	– 2 hours	Tokyo, Japan	+ 9 hours
Paris, France	+ 1 hour	Brisbane, Australia	+ 10 hours
Madrid, Spain	+ 1 hour	Sydney, Australia	+ 10 hours
Athens, Greece	+ 2 hours	Wellington, New Zealand	+ 12 hours

11   **a** If the Edinburgh time is 1400, what time is it in New York?

    b If it is 2315 in Edinburgh, what time is it in Athens?

    c If it is 1625 in Edinburgh, what time is it in Brisbane?

12   **a** If it is 1400 in Chicago, what is the time in Beijing?

    b If it is 1915 in New Delhi, what time is it in Paris?

    c If it is 0800 in Madrid, what time is it in Lahore?

> **Hint**   You may wish to convert to GMT first and then to the new time zone.

13 The opening ceremony of the 2016 Olympics took place on 5 August in the Maracana Stadium in Rio de Janeiro, Brazil.

The ceremony started at 2000 hours local time.

The opening ceremony was shown around the world.

What was the local time in these cities when the ceremony started?

  a Mexico City   **b** Moscow   **c** Tokyo   **d** Wellington

14 The 2017 IAAF World Championships took place in London. The opening ceremony took place on 4 August and began at 6 pm.

What was the local time in these cities?

  a Denver   **b** Sydney   **c** Hong Kong   **d** Rio de Janeiro

15 Mickie is flying from London to Orlando on holiday.

His flight leaves at 1520 GMT.

Orlando is GMT minus 5 hours. The flight takes 9 hours 15 minutes.

He reckons it will take 40 minutes to clear customs.

What local time should Mickie expect to leave Orlando Airport?

20 Considering the effects of tolerance

Exercise 20A Calculate limits and consider implications for compatibility

4 **Example 20.1**

A computer monitor is rectangular, with a length of 35 cm and breadth of 25 cm.

Each measurement has a tolerance of ± 0·5 cm.

a What is the lower limit of the length?

b What is the lower limit of the breadth?

c What is the least possible area of the monitor?

d Compare this with the nominal area.

a Lower limit (length) = 35 − 0·5 = 34·5 cm

b Lower limit (breadth) = 25·5 − 0·5 = 24·5 cm

c Least possible area = $l \times b$ = 34·5 × 24·5 ●————— Use the lower limits for length and breadth to work out the least possible area.

   = 845·25 cm²

d Nominal area = 35 × 25 = 875 cm²

   875 − 845·25 = 29·75 cm² ●————— Work out the difference.

The least possible area is 29·75 cm² smaller than the nominal area.

N4 **1** In an exam, pupils had to measure an angle.

The marking instructions stated: 'Accept an answer of (43 ± 2)°'.

Ten pupils gave answers as shown in the table.

Which pupils would be awarded the mark?

Pupil	A	B	C	D	E	F	G	H	I	J
Answer (°)	43	41	39	44	45	44	40	46	43	34

2 In the same exam, pupils had to draw a line 27 mm long.

The marking instructions stated: 'Accept an answer of 27 ± 2 mm'.

Ten pupils gave answers as shown in the table.

a Which pupils would be awarded the mark?

b What percentage of pupils were within tolerance?

Pupil	A	B	C	D	E	F	G	H	I	J
Answer (mm)	26	28	30	25	24	23	27	26	31	25

3 The tolerance of a machine part is set at 5·13 ± 0·05 cm.

 a What is the lower acceptable length for the part?

 b What is the upper acceptable length for the part?

 c What is the acceptable range for the part?

4 A part for a vacuum cleaner is set at 6·75 ± 0·15 cm.

 a What is the lower limit for accepting the part?

 b What is the upper limit for accepting the part?

 c What is the range of acceptable measurements?

5 A packet of sweets should contain 42 ± 0·5 g.

 Which of the following would meet this tolerance?

   42·3 g, 41·7 g, 41·4 g, 43·1 g, 42·7 g, 41·9 g, 42·6 g

  6 A watch-making company orders Lithium CR 2025 batteries for its watches.

 The height of the battery must be 2·5 mm to the nearest 0·05 mm.

 A quality control inspector checks a sample from a recent delivery.

 If more than 20% are outwith the acceptable tolerance, the delivery will be sent back.

 The sample results are shown below.

2·46	2·45	2·54	2·51	2·56	2·46	2·50	2·49	2·46	2·47
2·51	2·52	2·43	2·53	2·50	2·51	2·48	2·51	2·47	2·50

 Will the delivery be accepted?

 Use your working to justify your answer.

 7 A factory needs to order machine parts of size 45 mm.

 A tolerance of ± 2 mm is acceptable.

 The factory requires an accuracy rate of 85% or better.

 Two companies send samples of machine parts as shown below.

Company A (mm)	43	45	42	47	48	46	43	44
Company B (mm)	44	45	43	42	46	46	44	45

 From which of the two companies should the factory order?

 Use your working to justify your answer.

8 The diagram shows a child's construction brick and its dimensions.

 Each dimension is accurate to ± 0·06 cm.

 a Holly makes a wall which is 15 bricks long.

  What is the maximum length of Holly's wall?

 b Ellis makes a wall which is 6 bricks high.

  What is the minimum height of Ellis's wall?

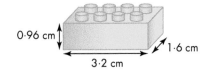

0·96 cm

3·2 cm

1·6 cm

9 Dawn has a set of 20 encyclopaedias, each of which is 32 ± 0·5 mm wide.

She has a book shelf which is 65 cm wide.

Can Dawn guarantee that all 20 encyclopaedias will fit the shelf?

Use your working to justify your answer.

10 A builder wants to fit a cement lintel above a set of windows.

The lintel is 3·6 ± 0·02 m long.

The space between the joists above the window is 3·7 ± 0·1 m.

Will the lintel always fit in the space above the window?

Use your working to justify your answer.

lintel

11 An engineering company decides to order nuts and bolts with screw diameter 7·5 ± 0·1 mm.

Will the nuts and bolts always fit together?

Use your working to justify your answer.

7·5 ± 0·1 mm

7·5 ± 0·1 mm

12 In the London 2012 Olympics, the splits of the US team who won the women's 4 × 100 m sprint relay were:

Tianna Bartoletta 11·12 s, Allyson Felix 9·97 s, Bianca Knight 10·33 s, Carmelita Jeter 9·70 s

The stopwatch was accurate to 0·005 of a second for each runner.

What would be the fastest time the team could record for the total race?

13 In the Olympic 4 × 100 m men's freestyle swimming competition in 2016 in Rio de Janeiro, gold and silver places were won by USA and France as shown.

Gold: United States	3:09·92	Silver: France	3:10·53
Caeleb Dressel	48·10 s	Mehdy Metealla	48·08 s
Michael Phelps	47·12 s	Fabien Gilot	48·20 s
Ryan Held	47·73 s	Florent Manaudou	47·14 s
Nathan Adrian	46·97 s	Jeremy Stravius	47·11 s

The stopwatch was accurate to 0·005 seconds for each swimmer.

a What is the slowest total time USA could have recorded?

b What is the fastest total time France could have recorded?

c Would the result be changed if the USA's slowest time and France's fastest time were used?

21 Investigating a situation involving gradient

Exercise 21A Using vertical and horizontal distances

N4 **1** Calculate the gradient of this car-port roof.

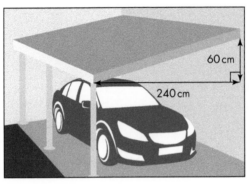

60 cm

240 cm

Hint Remember, the rule for finding the gradient of a slope is:

$$\text{Gradient} = \frac{\text{vertical distance}}{\text{horizontal distance}} = \frac{V}{H}$$

So the gradient of the slope

shown below $= \frac{60}{360} = \frac{1}{6}$

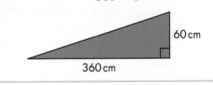

60 cm

360 cm

2 Calculate the slope of this ramp.

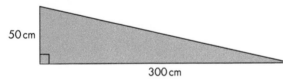

50 cm

300 cm

3 A surveyor is checking the slope of a field before building work starts.

She wants the slope to be less than $\frac{1}{100}$ to minimise preparation work. Does this field meet the requirements?

Use your working to justify your answer.

20 cm

24 m

4 Calculate the gradient of this Olympic ski-jump.

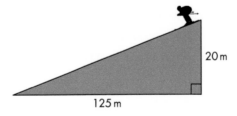

20 m

125 m

Example 21.1

Calculate the gradient of this slope.

Write your answer as a percentage.

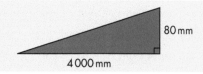

Gradient = $\dfrac{V}{H}$

$\qquad = \dfrac{80}{4000}$

$\qquad = \dfrac{1}{50}$

$\dfrac{1}{50} \times 100 = 2\%$ •⸺⸺⸺(To convert a fraction to a percentage, multiply by 100.)

5 Building regulations state that an access ramp must have a gradient of less than 1 in 12.

a Does this ramp meet building regulations?

Use your working to justify your answer.

b Calculate the length of the ramp.

6 An escalator is being installed to join the first and second floors of a shopping centre as shown.

For safety reasons, the slope should not be more than 1·5.

The company installing the escalator claims that it meets the safety regulations.

Is this claim valid?

You must show working to justify your answer.

6·1 m

3·9 m

7 A swimming pool in a leisure centre is 25 m long.

The shallow end is 0·9 m deep. The deep end is 2·4 m deep.

The slope of the bottom of the pool should be 0·07 ± 0·005.

Does the bottom of the pool fit these requirements?

You must use your working to justify your answer.

Hint | Use your knowledge of **tolerance** to help you answer this question.

8 A glider is coming in on a flight path which should have a gradient of 0·08 ± 0·005.

A glider is 3 km away from the airport, at a height of 235 m.

An air traffic control tells the glider pilot if he is too high, too low, or on track.

What advice will the air traffic control give the glider pilot?

Show all your working and justify your answer.

9 A public library is having an access ramp built.

Building regulations state:

- the maximum height of the ramp shall be 650 mm
- the maximum gradient of the ramp shall be 1 in 12.

Does the ramp shown meet these regulations?

Use your working to justify your answer.

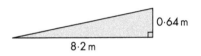

0.64 m

8.2 m

10 Ski slopes are graded partly due to their gradient.

The table shows the gradings.

Rating	Symbol	Grading	Gradient
Green circle	●	Easy	6–25%
Blue square	■	Intermediate	25–40%
Black diamond	◆	Advanced	Over 40%

For the following ski slopes, state in which category they lie.

a Horizontal distance 300 m, vertical distance 60 m

b Horizontal distance 125 m, vertical distance 75 m

c Horizontal distance 240 m, vertical distance 80 m

d Horizontal distance 272 m, vertical distance 68 m

11 In the United States, the grade of a road is its gradient written as a percentage.

A warning sign must be placed at the start of a section of road if:

- the gradient is more than 8%, and
- the section of road is more than 750 ft long.

A road rises 60 ft over a horizontal distance of 850 ft.

Does a warning sign need to be placed at the start of the section?

Show all your working and justify your answer.

5 **Example 21.2**

Find the gradient of the line joining the points $M(1, 2)$ and $N(5, 5)$.

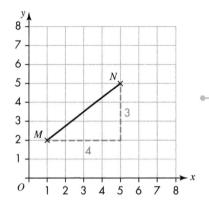

Plot the points M and N on a coordinate grid, and join using a straight line. Then form a right-angled triangle, shown with dotted lines in diagram. Count the vertical squares and the horizontal squares.

Gradient $= \dfrac{V}{H} = \dfrac{3}{4}$

Substitute the values into the formula to find the gradient of the line.

N5 **1** Plot these pairs of points.

Use the formula for calculating gradient to find the gradient between each pair of points.

a $A(3, 2)$ and $B(7, 5)$ b $C(1, 7)$ and $D(4, 3)$

c $E(6, 4)$ and $F(12, 4)$ d $G(3, 2)$ and $H(3, -1)$

2 Hyper couriers deliver packages around a town centre.

The amount charged per packet depends on the weight.

Hyper has three bands for charging, as shown in the table.

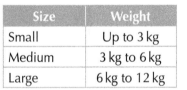

Size	Weight
Small	Up to 3 kg
Medium	3 kg to 6 kg
Large	6 kg to 12 kg

The graph shows how the cost changes within each band.

a Use the graph to state how much Hyper would charge to deliver these packets:

  i packet A, weighing 1 kg

  ii packet B, weighing 7 kg

  iii packet C, weighing 10 kg

b Calculate the gradient of each charging band.

c Which band (size of packet) gives the largest gradient?

d Why do you think that Hyper charges different rates for different sizes of packet?

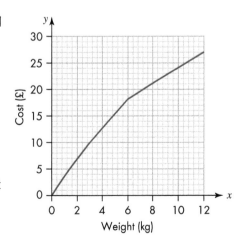

22 Solving a problem involving a composite shape

Exercise 22A Finding the area of a composite shape

N3 **1** The landing pit for a long jump event is a rectangle with dimensions as shown.

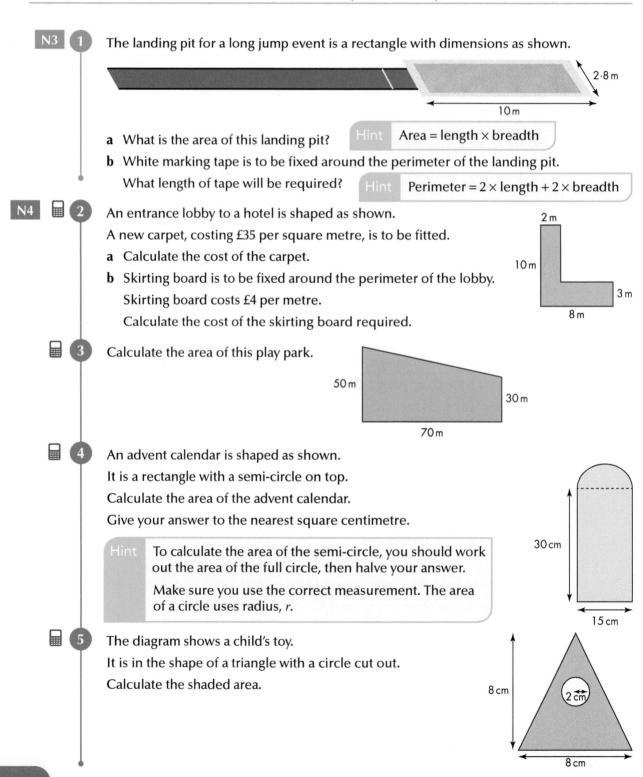

2·8 m

10 m

a What is the area of this landing pit? **Hint** Area = length × breadth

b White marking tape is to be fixed around the perimeter of the landing pit.
What length of tape will be required? **Hint** Perimeter = 2 × length + 2 × breadth

N4 **2** An entrance lobby to a hotel is shaped as shown.
A new carpet, costing £35 per square metre, is to be fitted.

2 m

10 m

3 m

8 m

a Calculate the cost of the carpet.
b Skirting board is to be fixed around the perimeter of the lobby.
Skirting board costs £4 per metre.
Calculate the cost of the skirting board required.

3 Calculate the area of this play park.

50 m

30 m

70 m

4 An advent calendar is shaped as shown.
It is a rectangle with a semi-circle on top.
Calculate the area of the advent calendar.
Give your answer to the nearest square centimetre.

30 cm

15 cm

Hint To calculate the area of the semi-circle, you should work out the area of the full circle, then halve your answer.

Make sure you use the correct measurement. The area of a circle uses radius, *r*.

5 The diagram shows a child's toy.
It is in the shape of a triangle with a circle cut out.
Calculate the shaded area.

8 cm

2 cm

8 cm

6 The Strictly Come Waltzing Dance Studio is re-laying its dance floor.

Anita has sketched the floor plan, which includes a semi-circular stage area.

a Calculate the area of floor to be re-laid.

b Wood flooring comes in packs of 4 m² costing £28·70 each.

How much will it cost to re-lay the dance floor, if they do it as efficiently as possible?

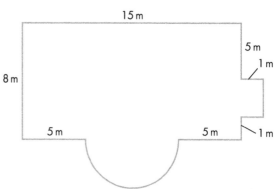

7 A landscape gardener has designed the garden below.

The length of the garden is 25 m and the breadth is 12 m.

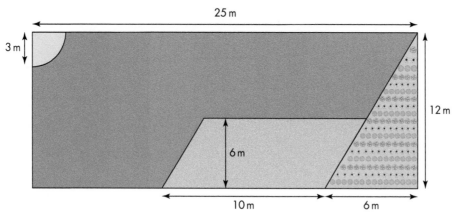

The landscape gardener has placed the following features in the rectangular garden:

• a small pond in the shape of a $\frac{1}{4}$ circle, radius 3 m

• a triangular flower area, with base of 6 m

• a patio area in the shape of a parallelogram, with base 10 m and height 6 m.

The area left is to be turfed.

What area of the garden is to be covered in turf?

Give your answer to **2 significant figures**.

8 An ornamental entrance is being built. The arch is semi-circular.

Both the front and back of the entrance need to be painted.

A 5-litre tin of paint covers 20 m².

Will one tin of paint be enough?

Use your working to justify your answer.

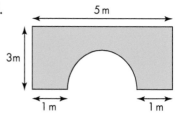

23 Solving a problem involving the volume of a composite solid

Exercise 23A Finding the volume of a composite solid

N3 **1** Calculate the volume of this cuboid made of wooden cubes.

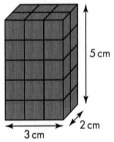

5 cm

2 cm

3 cm

2 Calculate the volume of this tissue box.

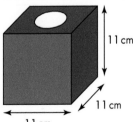

11 cm

11 cm

11 cm

3 Calculate the volume of this box of tea.

6 cm

5 cm

10 cm

4 Calculate the volume that this tea tray could hold.

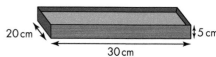

20 cm

30 cm

5 cm

5 The diagram shows a tent which is a triangular prism.

   a Calculate the area of the end.

   b Use your answer to part **a** to calculate the volume of the tent.

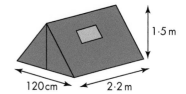

1·5 m

120 cm

2·2 m

6 The diagram shows a hexagonal prism.

The area of the base is 25 cm².

What is the volume of the prism?

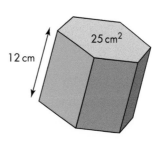

25 cm²

12 cm

7 The picture shows a tin of tuna.

The tin is a cylinder with dimensions as shown.

Calculate the volume of the tin.

6 cm

5 cm

TUNA

8 The picture shows a container of Camembert cheese.

The dimensions are as shown.

Calculate the volume of the Camembert.

Give your answer to **2 significant figures**.

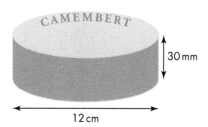

CAMEMBERT

30 mm

12 cm

9 The diagram shows a grain silo on a farm.

It is formed by a cylinder with a hemisphere on top.

Calculate the volume of the silo.

Give your answer to **2 significant figures.**

> **Hint** Calculate the volume of the cylinder and the volume of the hemisphere (which is $\frac{1}{2}$ of a sphere), using the formulae at the start of this exercise.
>
> Then add the two together.

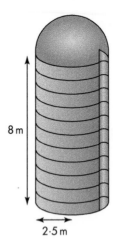

8 m

2·5 m

10 The picture shows a party hat in the shape of a cone.

Calculate the volume of the hat.

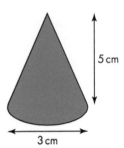

5 cm

3 cm

11 A golf ball has a diameter of 40 mm.

What is the volume of the golf ball?

Give your answer to **2 significant figures.**

40 mm

12 A monument is in the shape of a cuboid with a pyramid on top.

Its dimensions are as shown.

Calculate the volume of the monument.

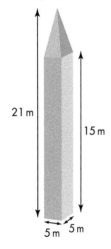

21 m

15 m

5 m 5 m

> **Hint** The formula for the volume V of a pyramid is:
>
> $V = \frac{1}{3} Ah$ where A is the area of the base and h is the vertical height.

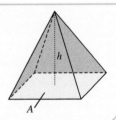

h

A

13 A 10-metre high diving platform is used in many diving competitions, including the Olympics.

The 10 m platform has dimensions as shown in the diagram.

What is the volume of concrete required to make this platform?

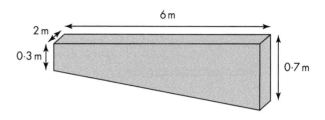

14 The diagram shows a foam tunnel in a child's soft-play area.

It is a cube with a cylindrical tunnel through it.

What volume of foam is used to make the tunnel?

Write your answer correct to **3 decimal places**.

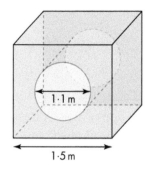

15 Kapat's Spice Company puts turmeric in cylindrical bottles with dimensions as shown.

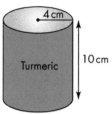

a What is the volume of the spice bottle?

b For ease of packing and transport, Kapat's decide to put the spice in a cuboid box.

The box has the same volume as the cylindrical bottle, and has a square base of 4 cm.

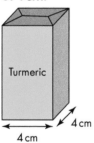

What is the height of the box?

24 Using Pythagoras' theorem

Exercise 24A Using Pythagoras' theorem in a two-stage calculation

N4 **Example 24.1**

Jenna is making a support for a shelf.

The horizontal shelf is 6 cm deep.

The support will fix to the wall 9 cm vertically below the shelf, as shown.

What is the length of support required?

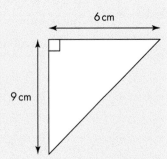

$a^2 + b^2 = c^2$ •————(Use Pythagoras' theorem.)

$\quad c^2 = 6^2 + 9^2$ •————(Rearrange with c^2 on the left-hand side, and substitute values for a and b.)

$\quad\quad = 117$

$\quad c = \sqrt{117}$

$\quad\quad = 10\cdot8\,\text{cm}$

N4 **1** Amber is putting together a bookcase.

The sides are vertical and the shelves are horizontal

She is going to cut a support strap to go diagonally across the back as shown.

How long should the support strap be?

Write your answer correct to 2 decimal places.

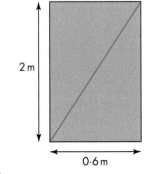

2 A ladder is resting against a vertical wall as shown.

What is the length of the ladder?

Write your answer to 2 significant figures.

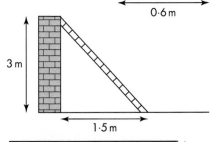

3 Janice is walking around the edge of a rectangular field.

How much distance would she save if she walked across the field diagonally?

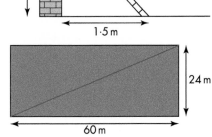

4 A 3 m ladder is placed up against a vertical wall.

The foot of the ladder is 0·4 m horizontally from the bottom of the wall.

How far up the wall does the ladder reach?

Give your answer to 2 decimal places.

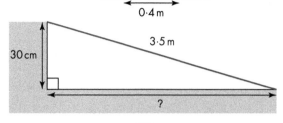

5 A metal ramp, 3·5 m long, is placed on a step 30 cm high.

How far away from the step does the foot of the ramp lie?

Write your answer correct to 1 decimal place.

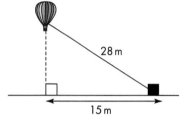

6 A hot air balloon is tethered to the ground with a rope 28 m long.

The tethering point is 15 m from a point vertically below the balloon.

How high is the balloon? Give your answer to 1 decimal place.

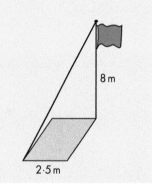

N5

Example 24.2

A flagpole, 8 m high, is fixed at a corner of a concrete square, of side 2·5 m, as shown.

A wire is fixed from the top of the flagpole to the corner of the concrete square to support the flagpole.

Calculate the length of the wire.

$a^2 + b^2 = c_1^2$ ●┄┄┄ (Step 1: Use Pythagoras' theorem to calculate the diagonal of the concrete square, c_1.)

$c_1^2 = 2·5^2 + 2·5^2$

$\quad = 12·5$ ●┄┄┄ (Don't take the square root, as you will square it again shortly.)

$c_2^2 = 8^2 + 12·5$ ●┄┄┄ (Step 2: Use Pythagoras again, but for a different triangle, where c_2 is the length of the wire, 12·5 is the square of the diagonal of the concrete square as previously calculated, and substitute the height of the flagpole.)

$\quad = 76·5$

$c = \sqrt{76·5}$

$\quad = 8·75\,m$

The wire is 8·75 m long.

N5 **7** A CCTV post is fixed to the corner of a concrete base as shown.

A metal support wire is fixed to the opposite corner of the base.

The post is 14 m high.

The concrete base is a square of side 3·2 m.

How long is the metal support wire?

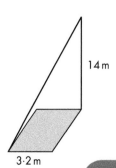

8 The internal dimensions of a container are:

length 2·3 m, height 2·2 m, breadth 2·1 m

Sally thinks she can fit her dinghy mast in the container if she places it along the space diagonally.

The mast is 3·7 m in length.

Is Sally correct?

Show all your working and justify your answer.

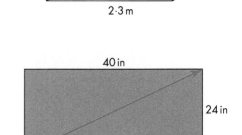

2·2 m
2·1 m
2·3 m

9 A TV is advertised as having a 46-inch screen.

A screen size is the diagonal across the face of the TV.

The length and breadth of the screen are measured as 40 inches and 24 inches as shown.

Is the claim of a 46-inch screen valid?

Show all of your working and justify your answer.

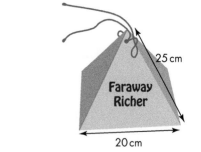

40 in
24 in

10 A Faraway Richer box of chocolates is in the shape of a square-based pyramid as shown.

The base has length of side 20 cm.

The sloping edge of the box is 25 cm long.

What is the height of the box?

Give your answer correct to 2 significant figures.

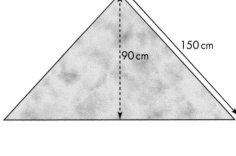

25 cm
Faraway Richer
20 cm

11 Angus has made a shawl in the shape of an isosceles triangle as shown.

He wants to put a fabric strip around the edge of the shawl.

He has 550 cm of fabric strip.

Is this enough?

You must show your working to justify your answer.

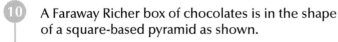

90 cm
150 cm

12 A sail, on a dinghy, is made up of two right-angled triangles as shown.

a Calculate the length of *AB*.

b Calculate the total area of the sail.

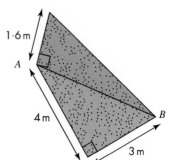

1·6 m
A
4 m
3 m
B

25 Extracting and interpreting data from different graphical forms

Exercise 25A Extracting and interpreting information in tables, charts, graphs and diagrams

N3 **1** Mr Nimmo is looking at the cost of hotels for a holiday.

A travel brochure shows the following table.

Hotel	1 week		2 weeks	
	Half board	Full board	Half board	Full board
Crown	£310	£390	£460	£520
Sceptre	£325	£405	£475	£535
Throne	£317	£397	£467	£527

How much will it cost Mr Nimmo to stay at:

a the Crown Hotel for 1 week, half board

b the Throne Hotel for 2 weeks, full board

c the Sceptre Hotel for 1 week, half board?

2 A class of S1 pupils were asked which social media site they used the most often.

The results are shown in the table.

Social media site	Facebook	Twitter	Instagram	WhatsApp	Flickr
Frequency	12	8	6	3	3

Draw and label a bar graph to show this information.

3 Caroline works in a car sales warehouse. The graph shows her car sales for one week.

a How many cars did Caroline sell on Tuesday?

b On which day did she sell the least number of cars?

c How many cars did Caroline sell in total?

d On which day did Caroline sell 9 cars?

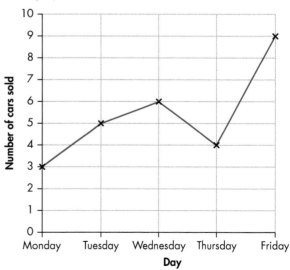

The table shows the results for a group of pupils who sat a maths test and a science test.

Pupil	A	B	C	D	E	F	G	H
Maths	55	73	52	62	25	67	95	35
Science	63	68	46	68	36	73	89	32

a Draw a scatter graph to illustrate this information.

b Draw a best-fitting line for your points.

c A pupil scored 60 marks in the maths test, but was absent for the science test.

Use your best-fitting line to estimate their likely score for the science test.

5 Traffic police noted the speed of cars, in miles per hour, as they drove along a stretch of road. The results were as follows:

31	52	43	49	36	35	33	29
54	43	44	46	42	39	55	48

a Construct an ordered stem and leaf diagram to illustrate this information.

b Due to road works the speed limit on the road was reduced to 50 mph.

How many cars would exceed this speed limit?

6 The table shows the cost to road tax a petrol or diesel car which was registered between March 2001 and March 2017.

Band	CO_2 emissions (g/km)	Single 12 month payment	Single 12 month payment by DD	Total of 12 monthly instalments by DD	Single 6 month payment	Single 6 month payment by DD
A	Up to 100	£0	N/A	N/A	N/A	N/A
B	101–110	£20	£20	£21	N/A	N/A
C	111–120	£30	£30	£31·50	N/A	N/A
D	121–130	£115	£115	£120·75	£63·25	£60·38
E	131–140	£135	£135	£141·75	£74·25	£70·88
F	141–150	£150	£150	£157·50	£82·50	£78·75

DD = direct debit   *Source: www.gov.uk*

a How much would it cost to road tax a car in band D for 6 months using direct debit?

b How much would it cost to road tax a car which has CO_2 emissions of 105 g/km if paying one single 12-month payment?

c How much would it cost to road tax a car which has CO_2 emissions of 138 g/km if it is paid monthly, over 12 months, by direct debit?

d John's car is in band D.

How much would he save by paying a single 12-month payment instead of two single 6-month payments, not using direct debit?

Times, in seconds, were recorded of men doing a 100 m sprint. The results were:

| 10·8 | 10·5 | 11·3 | 12·6 | 13·1 | 11·1 | 11·2 | 11·7 | 11·2 | 13·0 |
| 12·3 | 11·8 | 11·5 | 10·9 | 10·7 | 11·0 | 11·1 | 11·1 | 12·1 | 14·3 |

Times, in seconds, were also recorded for women doing a 100 m sprint.
The results were:

| 11·5 | 12·3 | 15·0 | 14·7 | 13·0 | 13·4 | 12·9 | 11·9 | 12·5 | 12·6 |
| 12·5 | 14·4 | 13·0 | 13·1 | 13·8 | 12·9 | 13·8 | 12·5 | 12·7 | 13·8 |

a Construct a back-to-back, ordered, stem and leaf diagram to illustrate this information.

b What is the median time for each group?

c What is the interquartile range for each group?

d Make one valid comment to compare the times recorded.

8 A manager in a large call centre recorded how many operators were absent each day over a 3-week period. The results were:

| 17 | 21 | 20 | 22 | 19 | 22 | 19 | 19 | 17 | 18 | 19 |
| 23 | 21 | 20 | 18 | 18 | 19 | 20 | 16 | 21 | 19 |

a Construct an ordered stem and leaf diagram to illustrate this information.

b What is the median of these data?

c What is the semi-interquartile range?

d If more than 20 operators are absent on a given day, the manager will have to call in agency staff.

What is the probability that on a day picked at random, the manager will have to call in agency staff?

9 Two mobile phone companies offer the following rates.

Company	Monthly charge (£)	Calls charge per minute (£)
On the Move	20	0·03
Under Cut	0	0·07

a Copy and complete the table showing the cost of making calls with each company.

Number of calls	0	200	400	1000	1500
On the Move cost (£)	20	26			
Under Cut cost (£)	0		28		

b Draw graphs for both companies on one set of axes.

c How many minutes of calls cost the same for each company?

d If Jennifer expects to make 750 minutes of calls, which company should she go with?

26 Making and justifying decisions using evidence from the interpretation of data

Exercise 26A Making and justifying decisions based on patterns, trends or relationships in data 🖩

N4 **1** The top two pupils in a class compare their marks for a set of history tests they did.

The results are shown in the table.

Test	A	B	C	D
Zainab	82	46	63	58
Torvil	78	50	57	60

The history prize goes to the pupil with the higher mean mark.

Who gets the prize?

2 Ben wants to get his car serviced.

He looks at the recent fixed costs, not including fixing faults, of services from two local garages.

Toolkit	£130	£187	£153	£174	£180	£142
Spark Plug	£167	£154	£191	£178	£125	£137

Ben decides to go to the garage with the lower mean cost.

Which garage should Ben go to?

N5 **3** The table shows the carbon emissions for different size cars.

Type of car	Emissions (kg/km)
Small petrol (up to 1·4 l)	0·18
Medium petrol (1·5–2·0 l)	0·22
Large petrol (> 2 l)	0·30
Small diesel (up to 1·7 l)	0·15
Medium diesel (1·8–2·0 l)	0·19
Large diesel (> 2 l)	0·26

For each type of car, calculate the emissions for a journey of 300 km.

Put the cars in order from 'most green' to 'least green'.

4 In 1954 Dr Roger Bannister broke the 4-minute mile barrier. Since then, the time to run the mile has gradually decreased.

The table shows the following 20 years of records being broken.

Plot these times on a graph.

Use the graph to predict when the '3-minute 30-second mile' could be run.

Date	Record time
06.05.1954	3 min 59·4 s
19.07.1957	3 min 57·2 s
06.08.1958	3 min 54·5 s
17.11.1964	3 min 54·1 s
17.07.1966	3 min 51·3 s
23.06.1967	3 min 51·1 s
12.08.1975	3 min 49·4 s

Exercise 26B Understanding the effects of bias and sample size

1 Brit-Rail kept a record of how many of their Edinburgh to Glasgow trains were late over a 2-month period. The results are shown in the table below.

January	1	0	2	7	3	1	2	4	5	1	2	4	7	4	3	0	1
	4	1	3	2	1	2	3	4	5	6	0	1	2	9			
February	5	6	5	4	3	7	8	1	4	3	4	5	6	4	3	4	2
	6	4	8	9	5	3	4	3	5	4	4						

a In which month was the more punctual service provided?

b Why is the amount of data different for each month?

c If January had the same number of days as February would your answer to part **a** be different?

2 A company has employees falling into one of three categories, as shown.

A survey to fairly represent the views of the employees is to be carried out.

a If one manager is to be included in the survey:

  i how many clerical staff should be included

  ii how many floor staff should be included?

Job title	Number of employees
Manager	7
Clerical	21
Floor staff	105

b If a sample size of 57 is to be taken, how many of each job title should there be?

3 A local council is doing a survey in a street. The street has houses numbered from 1 to 60, and bungalows numbered from 61 to 80.

Rather than ask residents in all the homes in the street, the council decides to take a sample of 24 homes.

For the survey to be unbiased:

a how many houses should be included in the survey

b how many bungalows should be included in the survey?

27 Making and justifying decisions based on probability

Exercise 27A Recognising and using patterns and trends to state the probability of an event happening

N3 **1** Caitlin tosses a coin.

What is the probability of it landing heads up?

Give your answer as a decimal fraction.

2 A teacher asked a class how they got to school and if they had remembered or forgotten their homework.

The results are shown in the table.

	Remembered homework	Forgot homework
Bus to school	15	5
Walk to school	6	4

What is the probability that a pupil picked at random:

a got the bus to school and had forgotten their homework?

Give your answer as a fraction.

b walked to school and had remembered their homework?

Give your answer as a percentage.

3 James spins a five-part spinner, like the one shown.

a Write, in words, the probability of James landing on a 1.

b Copy this probability scale and mark on it the probability of James landing on an even number.

N4 **4** A company produces car headlight bulbs.

The company is 97% confident that a bulb leaving the factory will work.

Halfiat's, a car parts company, buys a box of 200 headlight bulbs.

How many of the bulbs in the order are likely to not work?

5 A government report stated that 30 people died as a result of work-related accidents in 2016.

The same report stated that there were 3 400 000 people in work in 2016.

What is the probability that, in 2016, a worker picked at random was killed as a result of a work-related accident?

6 Beetown has a population of 7500.

The local hospital has indicated that a bird flu virus is spreading.

The chance of catching the virus in the first week is 0·02.

How many of the population are expected to catch the bird flu virus within the first week?

7 According to statistics there were 2·9 million vehicles registered in Scotland in 2015.

Police Scotland figures indicate that in 2015 there were 170 297 reported traffic offences.

What is the probability that in 2015 a driver of a registered vehicle, picked at random, would have a traffic offence held against them?

Exercise 27B Interpret and use probability to make and justify decisions

N5 **1** A school is running a fair to raise money for school funds.

One of the stalls advertises: 'Roll six 6s on a dice and win £1000'.

What is the probability of rolling six 6s?

2 Julie and Jeremy are playing a game which uses a dice and a spinner which is numbered from 1 to 8.

To win the game, Jeremy needs to get a 2 on the dice and a 2 on the spinner.

What is the probability of Jeremy getting the two 2s?

3 A street performer is doing a trick with a biased coin.

The probability of the coin landing heads up is 0·7.

If the performer tossed the coin 200 times, how many heads would she expect?

4 Dave takes the tram to work each morning Monday to Friday.

The probability of the tram being late is 30%.

How many times would Dave expect to be late in the course of 20 journeys to work?

5 A bag contains 30 beads.

The beads are in the ratio red:blue:yellow of 5:3:2.

Gemma picks one bead out at random and then replaces it.

She then picks out another bead at random.

What is the probability that Gemma picks out two beads of the same colour?

28 Preparation for assessment

Exercise 28A

  1 245 people took part in a 10-kilometre fun run.

$\frac{5}{7}$ of the runners finished in less than 1 hour.

How many runners finished in less than 1 hour?

 2 Two stretches of the Llangollen Canal are 15·3 km and 32·8 km long.

Gary has taken his narrowboat a distance of 39·7 km.

How far does he still need to go to complete these sections?

 3 A group of 180 S1 pupils were asked what their favourite subject was.

The results are shown in the pie chart.

How many pupils said maths was their favourite subject?

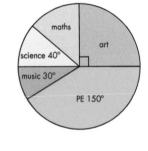

 4 200 cars were stopped at a road-side safety check by the police.

30% of the cars were found to have minor faults.

How many cars had minor faults?

 5 A kitchen breakfast bar is made from a semi-circle and rectangle as shown.

Calculate the area of the breakfast bar.

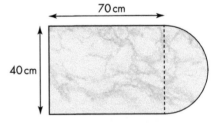

 6 St Cyrus and Inverbervie are two towns on the east coast of Scotland.

St Cyrus Tourist Office put out a leaflet with this claim of better weather:

> Over the last 10 years, St Cyrus has had an average rainfall of 53 cm.
>
> This is better than neighbouring Inverbervie, whose average rainfall is 54 cm.

The figures for rainfall, in centimetres, over the last 10 years for each town are shown in the table.

a For each town, calculate:

  i the mean **ii** the median.

b Make a valid comment on the claim made by St Cyrus Tourist Office.

St Cyrus	Inverbervie
50	60
52	52
57	52
53	64
51	52
65	51
56	58
53	40
58	54
53	57

7 Mike delivers car parts to garages in his van.

One day he took parts from his depot to a garage 180 km away.

The trip took him $2\frac{1}{2}$ hours.

a Calculate Mike's average speed for this journey.

b Due to road works, he could only average 60 km per hour on his return.

How much longer would it take Mike for his return journey?

8 A window cleaner has placed his ladder against a wall as shown.

Health and safety regulations state that the foot of the ladder should be at least 1 m from the base of the wall.

Is this ladder placed to satisfy the regulations?

Use your working to justify your answer.

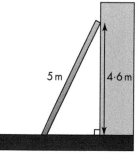

5 m   4·6 m

9 A storage company is packing equipment in containers.

A container can hold 10 kg.

There are four smaller boxes which weigh 7 kg, 8 kg, 2 kg and 3 kg.

Using a first-fit algorithm, how many containers will be needed to pack these boxes?

Show all your working and justify your answer.

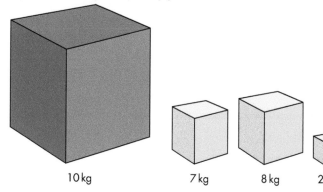

10 kg   7 kg   8 kg   2 kg   3 kg

N5 **10** Fred is a Quality Control Inspector in a factory which produces bags of flour.

One day he takes a sample of six bags of flour and records the weight, in grams, of each bag.

The results are: 400,   403,   400,   392,   403,   402

The mean of this sample is 400 g.

a Calculate the standard deviation for these data.

b Fred introduces new methods to improve consistency in weights.

He takes another sample and finds the mean to be 400 g and the standard deviation to be 5·8 g.

Are Fred's new methods successful?

Use your working to justify your answer.

11 Alpine Double Glazing firm offers their sales persons two options for pay.

Both options have a basic pay plus a commission on the monthly sales.

ALPINE	Option 1	Option 2
Monthly basic pay	£1 300	£975
Commission	0·7%	1·8%

One month, Eve sells £42 000 worth of double-glazed windows.

Which option would give her more pay for this month?

12 The diagram shows an ice rink used for curling.

The dimensions of the rectangle are: length 148 ± 2 ft, breadth 15 ± 0·7 ft.

The 'house' or 'target' consists of three rings with diameters of 4 ft, 8 ft and 12 ft.

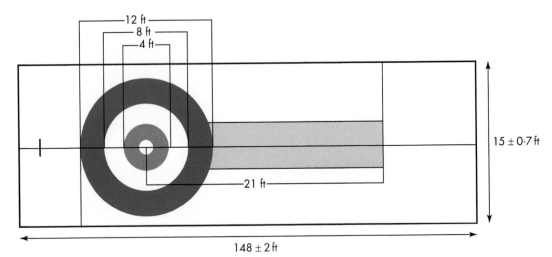

a What are the minimum and maximum lengths of the rectangle?

b The 'button', which is the centre circle, has a diameter of 1 foot.

What is the area of the red ring?

c James claims that the blue ring is twice the area of the white ring.

Is he correct?

Use your working to justify your answer.

d A guideline is being laid from the 'release point' to touch the outside ring as shown.

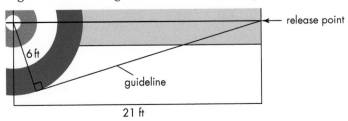

Calculate the length of this guideline.

13 The Edinburgh Dog and Cat Home has a number of dogs waiting to be adopted.

The weights, in kilograms, of the dogs are shown in the table.

Male	58	62	75	41	65	65	70	72	59
Female	69	51	70	48	65	55	65	68	49

a Construct a back-to-back stem and leaf diagram to illustrate this information.

b What is the weight of the heaviest female dog?

c What are the lower and upper quartiles of the weights of the female dogs?

d A boxplot is drawn for one set of dogs.

Does this boxplot represent the male or the female dogs?

e Construct a boxplot for the other set of data.

14 Gena's Gift Box Company makes gift boxes like the one shown.

It is in the shape of a cuboid with a lid formed from half a cylinder.

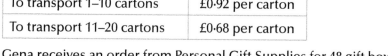

a Calculate the volume of the gift box.

b Gena packs the gift boxes into cartons for delivering to her outlet shop.

The cartons are cuboids measuring 48 cm in length, 20 cm in breadth, and 14 cm in height.

How many gift boxes can Gena fit into the carton?

c Gena works out her production costs as:

Item	Cost
To make one gift box	£4·23
To make one carton	£1·15
To transport 1–10 cartons	£0·92 per carton
To transport 11–20 cartons	£0·68 per carton

Gena receives an order from Personal Gift Supplies for 48 gift boxes.

Calculate how much it will cost Gena to make the gift boxes, pack them in cartons and transport them to the shop.

d Gena wants to make a profit of 30%.

What should she charge Personal Gift Supplies to cover her costs and make 30% profit?

e Personal Gift Supplies wants to make a profit of 20% on each box.

Calculate the minimum price to achieve this profit.

15 Lasswade High School's orienteering team are having a practice session.

The course they are running is:

- Start:
  - o run for 6 minutes at a speed of 75 metres per minute
  - o on a bearing of 150° to checkpoint 1
- Checkpoint 1:
  - o from checkpoint 1 run at 100 metres per minute for 8 minutes
  - o on a bearing of 025° to checkpoint 2
- Checkpoint 2:
  - o run directly from checkpoint 2 back to start.

a Using a suitable scale, construct a scale drawing of the orienteering course.

b Scott currently holds the school record, for the course, of 21 minutes 20 seconds.

One of the team doing the practice is running at an average speed of 5·1 km/h.

Will this beat Scott's record?

Use your working to justify your answer.

16 The table shows the exchange rate for £1 against the euro (€) and the South African rand (ZAR).

£1 (GBP)	€1·12
£1 (GBP)	17·65 ZAR

a Peter wants to buy a model car which is only available by ordering online from France or South Africa.

He looks at the costs of buying it from each country.

From France the cost will be €240·80.

From South Africa the cost will be 3 618·25 ZAR.

From which country should Peter buy the model car?

Show all your working to justify your answer.

b Peter then considers the cost of postage.

Sending the model car from France will cost €8·96.

Sending the model car from South Africa will cost 353 ZAR.

What impact will this have on Peter's choice of which country to buy from?

c For insurance purposes, Peter values the model car at £320.

He reckons it will appreciate in value at a rate of 8·2% per year.

What value should Peter insure it for in 3 years' time?

Round your answer correct to the nearest pound (£).